Cytokine Molecular Biology

A Practical Approach

Edited by

Fran Balkwill

ICRF Translational Oncology Laboratory,
St Bartholomews and the Royal London
School of Medicine and Dentistry,
Biological Sciences Building,
Charterhouse Square,
London EC1M 6BQ, U.K.

OXFORD
UNIVERSITY PRESS

OXFORD
UNIVERSITY PRESS

Great Clarendon Street, Oxford OX2 6DP

Oxford University Press is a department of the University of Oxford.
It furthers the University's objective of excellence in research,
scholarship, and education by publishing worldwide in

Oxford New York

Athens Auckland Bangkok Bogotá Buenos Aires Calcutta Cape Town
Chennai Dar es Salaam Delhi Florence Hong Kong Istanbul Karachi
Kuala Lumpur Madrid Melbourne Mexico City Mumbai Nairobi Paris
São Paulo Singapore Taipei Tokyo Toronto Warsaw

with associated companies in Berlin Ibadan

Oxford is a registered trade mark of Oxford University Press in the UK and
in certain other countries

Published in the United States by Oxford University Press Inc., New York

First edition published 1991
Second edition 1995
Reprinted 1997
Third edition 2000

Library of Congress Cataloguing in Publication Data
Cytokine molecular biology : a practical approach / edited by Fran Balkwill.—[3rd ed.].
(The practical approach series)
Includes bibliographical references and index.
1. Cytokines.   I. Balkwill, Frances R.   II. Series.
QR185.8.C95 C982 2000   616.07'9–dc21   00-057491
1 3 5 7 9 10 8 6 4 2

ISBN 0 19 963858 6 (Hbk)
ISBN 0 19 963857 8 (Pbk)

Typeset in Swift by Footnote Graphics, Warminster, Wilts
Printed in Great Britain on acid-free paper
by The Bath Press, Avon

Preface

During the past 10 years the study of cytokines has become central to biomedical research. Cytokines form a chemical signalling language in multicellular organisms that regulates development, tissue repair, haemopoiesis, inflammation, and the immune response. Potent cytokine polypeptides have pleiotropic activities and functional redundancy. They act in a complex network where one cytokine can influence the production of, and response to, many other cytokines. This bewildering array of effector molecules and associated cell-surface receptors has been simplified by the assignment of cytokines and their receptors into structural superfamilies; elucidation of convergent intracellular signalling pathways; and molecular genetics, especially targeted gene disruption to the 'knock-out' production of individual cytokines in mice.

It is also now clear that the pathophysiology of infectious, autoimmune, and malignant disease can be partially explained by the induction of cytokines, and that the subsequent cellular response and functional polymorphisms of cytokine genes contribute to the genetic programming of responses to pathogenic stimuli.

Cloning of cytokine and cytokine receptor genes has allowed the production of milligram quantities of purified protein for use in pre-clinical and clinical studies of acute and chronic infection, inflammatory disease, autoimmune disease, and cancer. Manipulation of the cytokine network with these recombinant proteins or other cytokine regulators provides a range of novel approaches to treating acute and chronic disease.

Cytokine Molecular Biology and *Cytokine Cell Biology* are the third editions of this popular book. The importance of the field is reflected in the expansion of the book into two volumes. Both volumes contain up-to-date methods for the study of cytokines, their receptors, and cytokine driven processes; the first focusing mainly on molecular biology, the second on cell-biology techniques.

The third editions of *Cytokine Molecular Biology* and *Cytokine Cell Biology* contain important new chapters, and revised and additional techniques have been added to the original chapters. Apart from the wide range of methods covered in the second edition, these two books also describe proteomics; conditional deletion of cytokine genes; measurement of intracellular signalling via the Jak/STAT and

MAPK pathways; purification of cytokine proteins; analysis of cytokine gene polymorphisms; intracellular fluorescent staining for cytokine protein; and biological assays for the newer cytokines.

With 216 protocols and 22 chapters in the two volumes, *Cytokine Molecular Biology* and *Cytokine Cell Biology* are not only essential for cytokine research, but are comprehensive guides to a wide range of techniques employed in biomedical research.

London   F.R.B.

September 2000

Contents

CONTENTS

Protocol list

Abbreviations

%CV	percentage variation of the mean
%II	percentage of the integrated intensity of the individual protein spot of the sum of the integrated intensities of all spots on the 2D gel
[^{3}H]TdR	tritiated thymidine
2-ME	2-mercaptoethanol
2D	two-dimensional
2DGE	2-dimensional gel electrophoresis
4-MUH	4-methylumbelliferyl heptanoate
A-SMase	acidic, endosomal sphingomyelinase
AET	aminoethylisothiouronium bromide (?hydrombromide?)
AML	acute myeloid leukaemia
AP	alkaline phosphatase
APAAP	anti-alkaline phosphatase
APC	antigen-presenting cell
APS	ammonium persulfate
ARA	American Rheumatism Association
ATCC	American Tissue (or Type) Culture Collection
BCDF	B-cell differentiation factor
BCG	bacille Calmette–Guérin
BCGF	B-cell growth factor
BDT	bis-diazotolidine
BFU-E	burst-forming units—erythroid colonies
BHK	baby hamster kidney
BLAST	basic local-alignment search tool
bo	bovine
Boc	*t*-butyloxycarbonyl
bp	base pair
BrdU	bromodeoxyuridine
BSA	bovine serum albumin
BSS	balanced salt solution
BTG	bovine thyroglobulin
c.p.m.	counts per minute
CAPK	ceramide-activated protein kinase

CAPP	protein phosphatase 2A
CD40L	CD40 ligand
CFC	colony-forming cells
CFU-E	colony-forming units—erythroid progenitors
CHAPS	(3-[(3 cholamidopropyl)dimethylammonio]-1-propane-sulfonate)
CHEF	contour-clamped horizontal electrical field
CHO	chinese hamster ovary (cells)
CID	collision-induced dissociation
cM	centimorgan
CM	conditioned medium
CMC	cell-mediated cytotoxicity
CMV	cytomegalovirus
ConA	concanavalin A
CPER	cytopathic effect reduction
CPG	controlled-pore glass
CPP32	(caspase 3)-like apoptotic protease
$CrCl_3 \cdot 6H_2O$	chromium chloride hexahydrate
Cre	*causes recombination*
CSF	colony-stimulating factor
CTL	cytotoxic T lymphocytes
CTSD	cathepsin D
CV	coefficient of variation
d.p.m.	disintegrations per minute
DAB	diaminobenzidine tetrahydrochloride
DAG	diacylglycerol
DEAE	diethylaminoethyl
DEPC	diethyl pyrocarobonate
dFCS	dialysed fetal calf serum
DGGE	denaturing gradient gel electrophoresis
DIC	differential interference contrast optics
DIG	digoxigenin
DME	Dulbecco's Modified Eagle's Medium
DMEM	Dulbecco's Modified Essential Medium
DMSO	dimethylsulfoxide
DR	HLA-DR alleles
DSS	disuccinimidyl suberate
DTT	dithiothreitol
E^-	non rosette-forming lymphocyte
E^+	E rosette-forming T cell
EBV	Epstein–Barr virus
EC	endothelial cells
ECGS	endothelial cell-growth supplement
EDA	eosinophil differentiation assay
EDMF	endothelial factor able to induce EC migration
EDTA	ethylenediaminetetraacetic acid
EGF	epidermal growth factor

EIA	enzyme immunoassays
ELAM-1	endothelial leucocyte adhesion molecule-1
ELISA	enzyme-linked immunoabsorbent assay
ELISpot	enzyme-linked immunospot assay
EMCV	encephalomycarditis virus
EMEM	Eagle's Minimum Essential Medium
EMSA	electrophoretic mobility-shift assay
Eo-CFC	eosinophil colony-forming cell
EP	eosinophil peroxidase
Epo	erythropoietin
EqS	equine serum
ERK 1,2	extracellular signal-regulated kinases
ES	embryonic stem cells
ESR	erythrocyte sedimentation rate
EST	expressed sequence tag
FACS	fluorescence-activated cell sorter
FADD	Fas-associated death domain
FALS	forward-angle light scatter
FAM	Carboxy-fluorescein
FBS	fetal bovine serum
FCA	Freund's 'complete' adjuvant
FCS	fetal calf serum
FDA	fluoroscein diacetate
FDG	fluorescein digalactosidase
FGF	fibroblast growth factor
FHS	Fischer's medium plus horse serum
FIA	Freund's 'incomplete' adjuvant
FITC	fluorescein isothiocyanate
Fmax	fluorescence at very high calcium
Fmin	fluorescence at very low calcium
fMLP	formylmethionyl-leucylphenylalanine
Fmoc	9-fluorenylmethoxycarbonyl
FRET	Förster resonance energy transfer
FSH	follicle-stimulating hormone
G-CFC	granulocyte colony-forming cells
G-CSF	granulocyte colony-stimulating factor
GAM	goat anti-mouse
GAPDH	glyceraldehyde phosphate dehydrogenase
GCP	granulocyte chemotactic protein
GFP	green fluorescent protein
GM-CFC	granulocyte and macrophage colony-forming cells
GM-CSF	granulocyte and macrophage colony-stimulating factor
Gp	G protein
GST	glutathione S-transferase
GuSCN	guanidinium thiocyanate
HBBS	Hanks' buffered salt solution

huCG	human chorionic gonadotropin
Hepes	4-(2-hydroxyethyl)-1-piperazine-ethanesulfonic acid
HFF	human foreskin fibroblast
HMP	4-hydroxymethyl-phenoxymethyl-copolystyrene
HPLC	high-performance (or pressure) liquid chromatography
HPTLC	high-performance thin-layer chromatography
HR	homologous recombination
HRP	horseradish peroxidase
HRR	haplotype relative-risk analysis
HS	horse serum
HSA	human serum albumin
HSV	herpes simplex virus
hu	human
huIFN	human IFN
huTNF	human TNF
huVEC	human umbilical venous endothelial cells
ICAM-1	intercellular adhesion molecule-1
ICCS	intracellular cytokine staining
IDDM	insulin-dependent diabetes mellitus
IEF	isoelectric focusing
IFN	interferon
IGIF	interferon-γ inducing factor
II	integrated intensity
IKKs	IkB kinases
IL	interleukin
IL-1ra	IL-1 receptor antagonist (product of *IL1RN*)
IL1RN	IL-1 receptor antagonist gene
IMDM	Iscove's Modified Dulbecco's Medium
iNOS	inducible nitric oxide synthase
IP	intraperitoneal
IP_3	inositol 1,4,5-trisphosphate
IPG	immobilized pH gradient
IPG	immobilized pH gradient
IRMA	immunoradiometric assay
ISH	*in situ* hybridization
IV	intravenous
Jak	Janus activated kinase
JNK	c-Jun N-terminal kinase
K_d	dissociation constant
KLH	keyhole limpet haemocyanin
KSR	kinase suppressor of Ras
LAB	labelled avidin–biotin
LAK	lymphokine-activated killer cell
LAL	Limulus amoebocyte lysate assay
LBD	ligand binding domain
LCR	locus control region

LDH	lactate dehydrogenase
LFA-1	lymphocyte function associated antigen 1
LGL	large granular lymphocyte
LIF	leukaemia inhibitory factor
*lox*P	*loc*us of *x*-ing over *P*1
LPBA	liquid-phase binding assays
LPS	bacterial lipopolysaccharide
LR	ligand/receptor
LSIMS	liquid secondary ion mass spectometry
LT	lymphotoxin
LTB4	leukotriene B4
LU	lytic units
M-CFC	macrophage colony-forming cell
M-CSF	macrophage colony-stimulating factor
mAb	monoclonal antibodies
MAC	monoclonal anti-cytokine
MACS	magnetic cell sorter
MAPK	p38 mitogen-activated protein kinase
MAPKAPK	MAPK-activated protein kinase
MAR	mouse anti-rabbit
MBP	myelin basic protein
MBS	maleimidobenzoic acid *N*-hydroxysuccinimide
MCCG	Monte Carlo composite genotype test
MCP	monocyte chemotactic protein
Meg-CFC	megakaryocyte colony-forming cell
MEM	Minimal Essential Medium
MF	mating factor
MFI	median fluorescence intensity
MFR	mannosyl fucosyl receptor
MHC	major histocompatibility complex
MKK	MAPK kinase
MKKK	MAPK kinase kinase
MLP	major late promoter
MMLV	Molony murine leukaemia virus
MMR	macrophage mannosyl receptor
MNC	mononuclear cells
MNPV	multiple nuclear polyhedrosis virus
Mø	macrophage
MOPS	3-*N*-morpholinopropanesulfonic acid
M_r	relative molecular weight
MS	mass spectroscopy
MTBE	methyl *t*-butyl ether
MTT	(3-[4,5-dimethylthiazol-2-ys]-2,5-diphenyl tetrazolium salt
mu	murine
muTNF	murine TNF
MW	molecular weight

MWCO	molecular weight cut-off
N-SMase	neutral plasma membrane-bound sphingomyelinase
NADG	N-acetyl-D-glucosamine
NBB	Naphthol Blue Black
NBS	N-bromosuccinimide
NBT	nitrobluetetrazolium
NCI	National Cancer Institute
NCS	newborn calf serum
NEPHGE	non-equilibrium pH gradient gel
NFκB	nuclear factor-κB
NGF	nerve growth factor
NGS	normal goat serum
NHS	normal human AB serum
NIBSC	National Institute for Biological Standards and Controls
NK	natural killer (cell)
NMMA	N^g-monomethylarginine
NMP	N-methylpyrrolidone
NMS	normal mouse serum
NPP	p-nitrophenylphosphate
NSD	neutral sphingomyelinase (activation) domain
NTA	nitrilotriacetic acid
OD	optical density
OLB	oligo-labelling buffer
OMP	orotidine monophosphate
OPD	orthophenylenediamine
ori	origin of replication
OVA	ovalbumin
PA	phosphatidic acid
PA-Ptase	phosphatidic acid-phosphatase
PAF	platelet activating factor
PAGE	polyacrylamide gel electrophoresis
PBA	phosphate-buffered saline + 0.1% bovine serum albumin
PBL	peripheral blood lymphocytes
PBMC	peripheral blood mononuclear cells
PBS	phosphate-buffered saline
PC	phosphatidylcholine
PC-PLC	phosphatidylcholine-specific phospholipase C
PC-PLD	phosphatidylcholine-specific phospholipase D
PCA	procoagulant activity
Pchol	phosphorylcholine
PCR	polymerase chain reaction
PdB	phorbol dibutyrate
PDGF	platelet-derived growth factor
PDQUEST	Protein Databases Quantitative Electrophoresis Standardized Test
PE	phycoerythrin
PEG	polyethylene glycol

PEth	phosphatidylethanol
PFA	paraformaldehyde
PFU	plaque-forming units
$PG1_2$	prostacyclin
PHA	phytohaemagglutinin
pI	isoelectric point
PI	phosphoinositol
PIP_2	phosphatidyl inositol bisphosphate
PK	proteinase K
PKA, -B, -C	protein kinases A, B, and C
PL	phospholipase
PLAP	placental alkaline phosphatase
PLC, -D	phospholipase C, -D
PMA	phorbol myristate acetate
PMC	2,2,5,7,8-pentamethylchroman-6-sulfonyl
PMEF	primary mouse embryo fibroblasts
PMN	polymorphonuclear cells
PMS	phenyl methezine sulphate
PMSF	phenylmethylsulfonyl fluoride
PmT	polyoma middle T
pNPP	p-nitrophenylphosphate
PPD	purified protein derivative
PrA	protein A
PrA-sRBC	PrA-conjugated sRBC
PRP	platelet-rich plasma
PTH	phenylthiohydantoin
PVDF	polyvinylidene difluoride
PVP	polyvinyl pyrrolidone
R	recombinant
ra	rat
RA	rheumatoid arthritis
RAC	rabbit anti-cytokine
RACE	rapid amplification of cDNA ends
RALS	right-angle light scatter
RAM	rabbit anti-mouse IgG
RANTES	regulated upon activation, normal T-cell expressed and secreted chemokine
RBC	red blood cells
RFLP	restriction fragment length polymorphism
RHPA	reverse haemolytic plaque assay
RIA	radioimmunoassay
RP-HPLC	reversed phase-HPLC
RPA	RNase protection assay
RR	NIBSC reference reagent
RT-PCR	reverse transcriptase polymerase chain reaction
SAC	staphylococcus aureus-Cowan

SAPK	stress-activated protein kinase
SC	subcutaneous
SCF	stem-cell factor
SD	standard deviation
SDS	sodium dodecyl sulfate
SDS-PAGE	sodium dodecyl sulfate-polyacrylamide gel electrophoresis
SEB	staphylococcal enterotoxin B
Sf	*Spodoptera frugiperda*
SF	synovial fluid
SFV	Semliki Forest virus
SG	specific gravity
SL	specific lysis
SLE	systemic lupus erythematosus
SM	sphingomyelin
SMase	sphingomyelinase
SPBA	solid-phase binding assays
sRBC	sheep red blood cells
SSC	saline–sodium citrate (buffer)
SSCP	single-stranded conformation polymorphism
STAT	signal transducer and activator of transcription
Sv	simianvirus
TAMRA	carboxy-tetramethyl-rhodamine
TBP	tributylphosphine
TBP	tributylphosphine
T_c	cytotoxic T cell
TCA	trichloroacetic acid
TCR	T-cell receptor complex
TDT	transmission disequilibrium test
TEMED	N,N,N',N'-tetramethylenediamine
TESPA	3-aminopropyltriethoxysilane
TET	tetra-chloro fluorescein
TFA	trifluoroacetic acid
TG	triglyceride
TGF-α, -β	transforming growth factor-α, -β
T_H	T-helper cell
T_{HP}	T-helper precursor cells
TIL	tumour-infiltrating lymphocytes
TLC	thin-layer chromatograhy
TMC	tonsillar mononuclear cells
TNF	tumour necrosis factor
TP	thyroid peroxidase
Tpo	thrombopoietin
TR	TNF receptor
TRADD	TNF-receptor associated death domain
TRAF	TNF receptor-associated factor
TRF	T-cell replacing factor

TT	tetanus toxoid (vaccine)
TUNEL	terminal transferase dUTP nick end labelling
UNG	uracil-DNA *N*-glycosylation
VCAM	vascular cell-adhesion molecule
VEGF	vascular endothelial growth factor
VNTR	variable number of tandem repeats
VSV	vesicular stomatitis virus
WGA-FITC	wheat-germ agglutinin–fluorescein isothiocyanate
YAC	yeast artificial chromosome

Chapter 1

Cloning and expression of cytokine genes

W. Declercq, M. Logghe, W. Fiers, and R. Contreras

Department of Molecular Biology, University of Ghent and Flanders Interuniversity Institute for Biotechnology (VIB), K. L. Ledeganckstraat 35, 9000 Ghent, Belgium

1 Introduction

Cytokines are proteins secreted into the extracellular fluid or are membrane exposed by a cell. There they exert their effects on the same cells (autocrine activity) or on neighbouring cells (paracrine activity) by interacting with specific receptors. Most cytokine proteins have, on a molar basis, very high biological activities (for instance, colony-stimulating factors are active at 10^{-11} to 10^{-13} M concentrations) which can usually be assayed in rapid, sensitive, and fairly specific *in vitro* test systems. These fast and simple detection systems have played a crucial role in cloning the corresponding cytokine genes by providing tools for either purification of the protein, the mRNA, or the direct detection of biologically active clones in cDNA expression libraries. Other cytokines, such as the strongly inducible chemokine family act in much higher concentrations (10^{-3} to 10^{-9} M) and have often been cloned by different techniques such as induction-specific differential hybridization or subtracted libraries. The mRNA of these molecules can represent 1% of the total poly(A)$^+$ RNA after stimulation. Recently the phenomenon of *in silico* cloning has been introduced, due to the fact that nowadays computer technology allows searches of protein and DNA databases to be made for sequence similarities (1, 2). In this way several related cytokines and cytokine receptors have been identified in recent years.

After cloning and sequencing the gene, efficient expression of the cDNAs has been obtained for most cytokines in prokaryotic, yeast or fungal, or mammalian systems. The most important steps concerning the cloning and expression of cytokine proteins, except for the techniques used for the purification of the natural or recombinant cytokine proteins, will be described in this chapter.

2 Synthesis of cDNA libraries

2.1 Preparation of mRNA

A prerequisite for the synthesis of a large, full-length cDNA library is an excellent preparation of mRNA. When handling RNA samples one should be extremely

1

careful to prevent RNase contamination. Several procedures for the preparation of eukaryotic total RNA have been described in detail in previous volumes of this series and in several laboratory manuals (3–6). A mammalian cell contains about 10 pg of RNA, of which 1–5% is mRNA. Rapid and efficient procedures, using oligo(dT$_{30}$) bound on latex (7) or magnetic beads (8), have been introduced and used successfully for the preparation of mRNA. The FastTrack 2.0 kit (Invitrogen) is routinely used in our laboratory. This method does not make use of RNase inhibitors as such, but instead uses a highly efficient protein/RNase degrader which yields intact, high-quality mRNA (see Section 2.2).

2.2 Conversion of mRNA to cDNA

The conversion of mRNA to cDNA has become a standard technique for which several commercial kits are available. In our laboratory we developed a protocol for cDNA synthesis and directional cloning based on the modified Gubler and Hofman method (3–5). Using this protocol we routinely obtain libraries containing about 10^6 to 10^7 independent clones (> 90% with insert) with an average insert length of ~ 1500 base pairs (bp). A detailed description is given in *Protocols 1* and *2* (see also *Figure 1* for the vectors used). The procedure can be modified according to the way the cDNA will be inserted into the vector. The choice of the vector, in turn, is dependent on the way the library will be screened (see Section 3).

Protocol 1
cDNA synthesis and cloning

Equipment and reagents

- 5 × first-strand reaction buffer: 250 mM Tris–HCl pH 8.3, 375 mM KCl, 15 mM MgCl$_2$ (Gibco BRL)
- 5 × second-strand reaction buffer: 100 mM Tris–HCl pH 6.9, 450 mM KCl, 23 mM MgCl$_2$, 0.75 mM β-NAD$^+$, 50 mM (NH$_4$)$_2$SO$_4$ (Gibco BRL)
- 10 × T4 ligation buffer: 660 mM Tris–HCl pH 7.5, 50 mM MgCl$_2$, 10 mM dithioerythritol, 10 mM ATP (Boehringer)
- Biogel column buffer: 10 mM Tris–HCl pH 7.5, 1 mM EDTA, 0.05% SDS (w/v)
- TE buffer: 10 mM Tris–HCl pH 7.5, 1 mM EDTA
- 1 M dithiothreitol (Gibco BRL)
- 1% NP-40 (w/v)
- 5 M NaCl

- 1% BSA (w/v)
- 1 M β-mercaptoethanol
- Phenol/chloroform/isoamyl alcohol (50:48:2 mixture by vol.)
- Diethyl ether
- Yeast tRNA (Sigma)
- 10 mM dNTP mix (Boehringer)
- [α-^{32}P]dCTP (3000 Ci/mmol, 10 mCi/ml; Amersham)
- 1 μg/μl *Not*I-primer-adaptor (Promega)
- 50 pmoles/μl phosphorylated *Sfi*I adaptors (5′-GTTGGCCTTTT-3′ and 5′-AGGCCAAC-3′)
- 40 U/μl RNase block ribonuclease inhibitor (Stratagene)
- 10 U/μl *E. coli* DNA ligase (Gibco BRL)

- 200 U/μl Superscript II reverse transcriptase (Gibco BRL)
- 1 U/μl T4 DNA ligase (Boehringer)
- 10 U/μl DNA polymerase I (Gibco BRL)
- 1 U/μl Klenow enzyme (sequencing grade; Boehringer)
- RNaseH (Gibco BRL)
- Sephacryl S-400 spun columns (Pharmacia)
- Biogel A150M column (∅ 0.7 cm, length 9 cm; Bio-Rad)
- Evaporator (Savant, SVC-100)

Method

1 Mix the mRNA (5 μg in a volume of 7.5 μl deionized water) and 2 μl of the *Not*I-primer-adaptor, heat to 70°C in a 1.5 ml centrifuge tube for 10 min, and immediately quench on ice (to denature the secondary structures).

2 In a separate centrifuge tube, mix the following reagents: 4 μl of the first-strand reaction buffer, 2 μl of 0.1 M dithiothreitol, 1 μl of a dNTP, and 0.5 μl of RNase block ribonuclease inhibitor. Add this mixture to the mRNA. After 2 minutes at 42°C (to equilibrate the temperature), add 3 μl of SuperscriptII reverse transcriptase and incubate the reaction at 42°C for 1 hour.

3 Put the reaction mixture on ice and add the following reagents in this order: 88 μl of sterile deionized water, 30 μl of second-strand buffer, 3 μl of dNTP mix, 3 μl of [α-^{32}P]dCTP, 1 μl of *E. coli* DNA ligase, 4 μl of DNA polymerase I, and 1.4 U of RNaseH. After 2 h (no more!) at 16°C, add 2 μl of Klenow enzyme and incubate the mixture at 16°C for 30 min.

4 Extract the reaction mixture once with phenol/chloroform/isoamyl alcohol and purify over two Sephacryl S-400 spun columns in parallel.

5 Add to the cDNA (pooled column effluent approximately 150 μl): 12 μl of phosphorylated *Sfi*I-adaptors, 20 μl of ligation buffer (Boehringer), 5 μl of 1% NP-40 (enhancer of blunt ligations), 3 μl of T4 DNA ligase, and deionized water to a total volume of 200 μl. Incubate this mixture for 16 h at 12°C.

6 Extract the reaction once with phenol/chloroform/isoamyl alcohol, three times with diethylether, and then evaporate for 15 min at 65°C.

7 Add 6 μl of 5 M NaCl (150 mM final concentration), 2 μl 1 M β-mercaptoethanol, and 2 μl of the 1% BSA solution to the cDNA mixture (total volume of 200 μl). Then add 40 U *Not*I and incubate for 3 h at 37°C. Concentrate by evaporation to a volume of about 100 μl.

8 Purify the radioactive cDNA and fractionate on a Biogel A150M column (max. loading volume 100 μl), previously saturated with 100 μg yeast tRNA and rinsed overnight with column buffer. Collect the cDNA fractions according to the counts in a scintillation counter: large cDNAs in the first fractions, smallest in the later ones. Extract the cDNA fractions (separate or pooled) with phenol/chloroform/isoamyl-alcohol, ethanol-precipitate in the presence of 10 μg of yeast tRNA and resuspend in TE buffer.

Protocol 1 continued

9 This cDNA can then be inserted in a directional way into an *Sfi*I–*Not*I opened cloning vector (see *Figure 1*). It is advisable to perform test ligations in which the cDNA and the vector are titrated. Usually approximately 1 µg of purified vector is used for pooled cDNA obtained from 5 µg of poly(A)$^+$ RNA. However, the cDNA is best ligated in 10–20 small ligations, rather than to scale-up the best test reaction.

10 Pool the ligation reaction mixtures, extract with phenol/chloroform/isoamyl alcohol (and wash the phenol phase to avoid losing too much DNA), and ethanol-precipitate with 10 µg of yeast carrier RNA. Finally dissolve in 20 µl of deionized water. The cDNA library is now ready for transformation to *E. coli*.

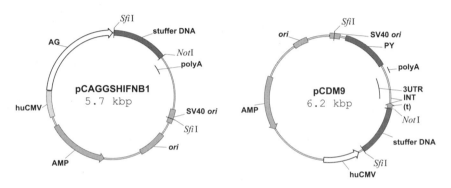

Figure 1 Structure of two mammalian expression vectors used for directional cDNA expression libraries. Both vectors are designed for high-level transient or stable expression of cloned *Sfi*I/*Not*I cDNA in mammalian cells. Preparation of the vectors for cDNA ligation is performed by a partial *Sfi*I/*Not*I digest because the SV40 *ori* contains an indispensable extra *Sfi*I site. These vectors contain: the β-lactamase gene for selection in *E. coli* (AMP), the pMB1 origin of replication from *E. coli* (*ori*), the SV40 fragment containing the bidirectional origin of replication (SV40 *ori*), and the human IFN-β gene which has as stuffer function (stuffer DNA). pCAGGSHIFNB1 carries a chicken β-actin/rabbit β-globin hybrid promoter (AG), the human cytomegalovirus immediate–early promoter enhancer sequence (huCMV), and the 3′ untranslated region of rabbit β-globin with inclusion of the (poly)A signal (polyA). In pCDM9, the protein encoded by the inserted cDNA can be expressed under control of the CMV promoter (huCMV) and the SV40 polyadenylation site (polyA). pCDM9 also carries polyomavirus DNA containing the polyoma origin of replication and enhancer (PY), and SV40 small T-antigen sequences: part of the small T-antigen (t), intron (INT), and 3′ untranslated region (3UTR). (pCAGGSHIFNB1: BCCM Acc. No. 2847; pCDM9: BCCM Acc. No. 2788; http://www.belspo.be/bccm/lmbp.htm)

Since the transformation frequencies of *E. coli* cells obtained by improved protocols and by rather simple electroporation procedures routinely vary between 10^8 and 10^9 transformants/µg DNA, it is possible to construct libraries in plasmids encompassing several million clones starting from 5 µg poly(A)$^+$ mRNA. A typical scheme is described in *Protocol 2*.

Protocol 2
Electrotransformation of *E. coli* cells

Equipment and reagents

- LB broth: 1% Bacto tryptone Difco (w/v), 0.5% Bacto yeast extract Difco (w/v), 0.5% NaCl (w/v)
- SOC broth: 2% Bacto tryptone Difco (w/v), 0.5% Bacto yeast extract Difco (w/v), 10 mM NaCl, 2.5 mM KCl, 10 mM MgCl$_2$, 10 mM MgSO$_4$, 20 mM glucose
- 10% ice-cold glycerol (v/v)
- Chilled centrifuge (Sorvall RC5B) and rotors (SLA-3000)
- Gene Pulser (Bio-Rad) and 0.2 cm gap electroporation cuvettes
- 17 × 100 mm round-bottomed tubes

Method

1 Grow an *E. coli* culture overnight to saturation (OD_{600} at least 2–5).

2 Inoculate 200 ml of LB broth with a 1/100 volume of the fresh overnight culture and grow cells with vigorous shaking (300 r.p.m.) to an OD_{600} no higher than 0.5 (~ 2 h).

3 Chill the flask on ice for 15–30 min, centrifuge in a cold rotor at 4000 g for 15 min and resuspend the cells in an equal volume of sterile ice-cold bidi. Centrifuge as before.

4 Resuspend in 100 ml of ice-cold bidi and centrifuge as in step 3.

5 Resuspend in 4 ml of ice-cold 10% glycerol. Centrifuge as in step 3, and resuspend in a final volume of 0.6 ml ice-cold 10% glycerol.

6 Use the cell suspension directly or snap-freeze in a dry-ice/ethanol bath in aliquots of 40 µl and store at −70°C. Note, however, that fresh, unfrozen cells will generate about five times more colonies for a given amount of plasmid DNA than frozen ones.

7 Mix 40 µl of the cell suspension with 2 µl of the cDNA library (see Protocol 1) in a chilled tube. Mix well and incubate on ice for about 1 min.

8 Transfer the mixture to a chilled 0.2 cm gap electroporation cuvette and apply an electric pulse (25 µF, 2.5 kV, 200 Ω).

9 Immediately add 1 ml of SOC broth (very important), transfer to a 17 × 100 mm round-bottomed tube and incubate at 37°C for 1 h under mild shaking (~ 200 r.p.m.) before plating.

3 Screening the cDNA libraries for cytokine genes

Historically, the first cytokine genes were isolated on the basis of biological signals obtained by injection of mRNA released from filters containing pools of cDNA plasmids (hybrid release selection) into *Xenopus laevis* oocytes (9–13). As this procedure requires an abundant source of highly active mRNA it has lost its popularity. Later on, several cytokine genes were isolated by hybridization selection with DNA probes deduced from the amino-acid sequence (14–18). In

5

addition, elegant selection strategies were used to isolate the genes for cyto-kines by direct expression in *in vitro* (19–21) or *in vivo* systems (22–24). More recently, cytokine genes in which the biological activity was identified after the clone was obtained have been isolated by differential screening strategies (25–28). Another 'blind' method that has led to the identification of novel secreted proteins in general, and to the isolation of new chemokines in particular, is the signal sequence trap method (29, 30). Nowadays many novel genes are identified on the basis of computer-assisted homology searches of public expressed sequence-tag (EST) databases (*in silico* cloning).

Screening procedures on the basis of biological activities are described in Section 3.1, screening procedures based on data obtained from the purified protein are described in Section 3.2, and screening on the basis of homology is mentioned in Section 3.3.

3.1 Screening on the basis of biological activity

3.1.1 Hybrid selection

The first cytokines were cloned on the basis of a selection procedure called 'hybrid selection' and injection of the hybridized mRNA fraction into X. *laevis* oocytes (9–13). This method is no longer commonly used. Nevertheless, we want to mention a variation to this technique established by Lomedico and colleagues (31). They used an elegant selection technique to isolate the cDNA clone for murine (mu) IL-1β based on the availability of a goat anti-muIL-1 antiserum. Groups of cDNA clones were screened by hybrid selection using mRNA derived from IL-1 producing cells. The released mRNA fractions were then translated in reticulocyte lysates, and the proteins were immunoprecipitated by means of the specific antiserum and analysed by polyacrylamide gel electrophoresis (PAGE). It was estimated that only 0.005% of the poly(A)$^+$ RNA from superinduced cells coded for this IL-1β precursor protein, showing the great sensitivity of this screening technique.

3.1.2 Direct expression by *in vitro* transcription

The simplicity and high efficiency of an *in vitro* expression system for the expression of a cytokine was first illustrated by injecting into X. *laevis* oocytes transcripts obtained from the human (hu) IFN-β gene transcribed from the phage P_L promoter by means of E. *coli* RNA polymerase (19). The cDNA clone for the B-cell growth factor muIL-4 was the first to be isolated using an *in vitro* selection technique based on the translation of SP6 polymerase-produced transcripts into X. *laevis* oocytes (32). Transcripts of the total cDNA library (45 000 clones) as well as an enriched library (4000 clones) produced, upon translation, a biological signal that allowed the isolation of an IL-4 clone. The same procedure was used to isolate the cDNA clone for IL-5 (33). Nowadays many commercial suppliers offer efficient *in vitro* coupled transcription/translation kits based on SP6, T3, or T7 RNA polymerase. Recently it was shown that such strategies could lead to the successful isolation of cDNA clones (34).

3.1.3 Direct expression in bacterial cells

Although the direct expression of eukaryotic cDNA in bacterial cells as such is practically impossible, eukaryotic coding sequences can be expressed in bacteria after fusion to bacterial proteins. This fusion protein can then be detected if an antibody is available. Immunological screening for proteins is carried out most frequently with recombinant expression libraries constructed in either plasmid or bacteriophage λgt11 vectors (35). Since phage λ-based vectors are more efficiently introduced into bacteria than plasmids and phage plaque screening is generally easier, we would advise the use of the latter system. Indeed, in this way a human, high molecular-weight B-cell growth factor was cloned from phyto-haemagglutinin (PHA)-stimulated Namalva cells using the λZAP (Stratagene) expression vector (36). As in many cases antibodies against the cytokine are not available, this cloning procedure has seldom been followed. However, several authors have reported on the development of chimeric proteins consisting of the extracellular region of cytokine receptors fused to the hinge region of human IgG1 or -3 Fc sequences and these have been successfully used as an alternative for ligand-specific antibodies in expression cloning strategies (24, 37–40). Therefore, the bacterial direct expression strategy together with the direct expression by *in vitro* transcription methodology (see Section 3.1.2) might be reconsidered as useful tools in the cloning of ligands of orphan receptors.

3.1.4 Direct expression in mammalian cells

The possibility of identifying mammalian genes by direct expression of their cDNAs in mammalian cell lines was first documented by Okayama and Berg in 1982 (41). Different mammalian expression vectors were developed. These vectors contain at least four basic elements:

- an efficient eukaryotic transcription unit (containing strong eukaryotic promoters such as the SV40 early or late promoters (42, 43), the cytomegalovirus early promoter (44), or the β-actin promoter (45);
- a splice junction;
- an origin of replication (*ori*) active in mammalian cells (mostly SV40- or polyomavirus-based *ori*);
- a prokaryotic *ori* and selection marker.

The first cytokine identified in a mammalian expression library was murine mast-cell growth factor, which is identical to muIL-3 (22). After selection of a number of possible candidates by the hybrid release technique (see Section 3.1.1), a full-size clone was isolated by transient expression of the plasmids in COS-7 monkey cells. COS-7 cells were used as the recipient for this plasmid, because the expressed SV40 large T antigen drives the replication of the expression plasmid from the SV40 *ori*. The transfected cells were able to express the interleukin to levels that were 300-fold above the background. The same group also used this technique for the selection of human, full-size GM-CSF cDNAs in a library of 10 000 individual colonies (23). A similar technique was applied by a group at the Genetics Institute to isolate the genes for huGM-CSF and huIL-3 (46, 47). The cloning of the latter proved to be a complicated task. Indeed, the usual technique of cross-hybridization between murine and human genes was inapplicable because of low nucleotide sequence homology (47). The problem was further complicated due to a low expression level of the cytokine by activated human peripheral blood lymphocytes. Hence, the multi-CSF (IL-3) clone was isolated from the gibbon cDNA library by screening transfected COS-1 cells for biological activity in the huGM-CSF assay in the presence of antiserum to GM-CSF. The human clone was then identified by cross-hybridization with the gibbon clone. More recently, the 4-1BB ligand (24), the Fas ligand (39), the pre-B-cell growth-stimulating factor/stromal cell-derived factor-1 (48), thrombopoietin (49), huIL-16 (50), and chicken IL-2 (51) have been cloned using direct expression technology.

Most of the direct expression-based strategies for cytokines rely on the availability of a sensitive biological test, often making use of a factor-dependent proliferation assay. At present, numerous receptor cDNAs have been cloned, and sequence comparison has allowed the identification of receptor families such as the TNF receptor and the chemokine receptor superfamilies (52, 53). In addition to this, similarity searches in EST databases and degenerate oligonucleotide-based RT-PCR has led to the cloning of numerous orphan receptors. Transient or stable transfection of orphan receptors in mammalian cells can then be used to establish biological assays for receptor signalling (17, 49, 54). Alternatively, especially in the case of membrane-bound ligands, artificially fusing the extracellular parts of membrane receptors to immunoglobulin Fc domains can provide powerful tools for isolating their respective ligands by means of direct expression cloning (24, 39, 55). As a consequence, these chimeric receptor–IgFc fusions can also be used to identify the best cellular source to start the cDNA cloning from, or to purify the cytokine of interest (see Section 3.2).

Besides transient expression systems, as described earlier, the potential use of more permanent expression systems for cloning new cytokines has been illustrated (56, 57). They synthesized a directional cDNA library in a retroviral vector. By functional screening in the appropriate cell lines they could then isolate the cDNAs for IL-3 and GM-CSF.

3.2 Screening on the basis of protein sequence data

When a portion of the amino-acid sequence of a protein is known, an oligonucleotide can be designed based on this information. However, due to the degeneracy of the genetic code, the selection of the protein region for reverse translation into a nucleic acid sequence is somewhat complicated. Several aspects, such as length, G + C content, self-complementarity, and complexity have to be considered. A detailed analysis of these considerations is described in the literature (58–60). The technique of oligonucleotide screening was used successfully for a number of lymphokines (e.g. huM-CSF (61), huG-CSF (15), muGM-CSF (62), huTNF (12), muTNF (13), huIL-6 (63), huIL-15 (64), and huIL-18 (65). In some cases, a special technique of overlapping oligonucleotide probes may be advantageous to obtain high-radioactive probes by a simple fill-in polymerization using Klenow enzyme (13). If sufficient protein sequence data are available (e.g. different fragments) one can try to obtain hybridization probes by making use of PCR technology. Therefore, degenerate sense and antisense oligonucleotide primers are designed on the basis of different amino-acid stretches (16, 65, 66). The interjacent sequences can then be amplified in a PCR reaction, generating larger probes with complete homology.

3.3 Screening on the basis of homology

Once the cytokine cDNA sequence for one organism (often human or mouse) has been cloned, it is often desirable to isolate the cDNA for the same gene function in other organisms to determine the degree of relatedness at the molecular level, functional cross-reactivity between species, etc. Heterologous, low-stringency hybridization techniques are often applied for this. Before screening heterologous libraries, the usefulness of the probe is tested on genomic DNA digests or mRNA preparations of cells that are known to express the gene of interest.

One can also use sequence data from known cytokines to isolate novel cytokines. Samal *et al.* designed a degenerate oligonucleotide on the basis of similarity in the coding sequences at the signal peptidase-processing site (67). This probe allowed the isolation of known cytokines, such as IL-1β, IL-6, IL-8, and the novel cytokine pre-B-cell colony-enhancing factor, by hybridization. A similar strategy has been applied to the cloning of cytokine receptors containing the conserved WSXWS motif (68). Hybridization probes can also be synthesized by PCR technology starting from oligonucleotides derived from more conserved regions in the proteins. In this way, additional clones can be obtained if the targeted protein belongs to a family of related cytokines or leads to the cloning of cytokines from other species. This strategy allowed the cloning of cytokines (69), cytokine receptors (70), as well as other proteins (71).

More recently, numerous EST sequence databases and computer applications to search for sequence similarities have become available to the molecular biologist (e.g. http://www.ncbi.nlm.nih.gov/BLAST). BLAST programs allow screening with both nucleotide or protein query sequences against nucleotide or protein sequence databases (1, 2). By making use of these tools it is possible to identify complete or partial cDNAs related to the protein of interest, thereby omitting the need to obtain a cDNA library (*in silico* cloning). In this way, several related cytokines cDNAs belonging to the tumour necrosis factor (TNF) ligand super-family and the chemokine family have been identified (72, 73; see also Section 3.2). Partial cDNAs can be further completed by means of 5'- and/or 3'-RACE. This methodology can also be applied in strategies mentioned in Section 3.2.

4 Heterologous expression of cytokine genes

After a putative cytokine cDNA clone has been isolated by a technique which is not based on direct expression, it is most important to demonstrate that the clone obtained can produce a biologically active cytokine. For this purpose, several expression systems are available. However, a mammalian system will often be chosen, because these systems guarantee a more faithful translation and processing from a minimally manipulated clone, although the expression levels are often rather low. Indeed, secondary modifications will be carried out and secretion will almost certainly be observed if the complete information for the signal sequence is present in the cDNA clone. Once a biologically active clone is identified, a strategy has to be selected for the preparative synthesis of the desired cytokine in larger quantities; four main options are available:

- expression in mammalian cells (see Section 4.1);
- expression in insect cells (see Section 4.2);
- expression in lower eukaryotes (see Section 4.3); or
- expression in bacteria (see Section 4.4).

4.1 Expression in mammalian cells

Mammalian expression has a number of advantages that arise from the fact that the heterologous gene is expressed in a host cell that most closely resembles its natural environment. The proteins fold in a correct way and they form the appropriate disulfide bridges. Furthermore, post-translational processing and modifications (especially glycosylation) will be more faithful. Two examples of mammalian expression systems will be briefly discussed: a transient expression system (see Section 4.1.1) and a permanent expression system (see Section 4.1.2).

4.1.1 Transient expression

Efficient vectors for transient expression in cells, based on SV40 replication and expression signals, were initially developed by Gheysen and Fiers (42). Currently, many companies sell optimized vectors using strong promoters such as SV40 early or late promoters, the cytomegalovirus early promoter, or the β-actin promoter. Transient expression of a gene is obtained by the transfection of an expression vector, mostly containing an SV40 *ori*, in host cells that express constitutively SV40 large-T antigen, such as COS or HEK293T cells. Large-T is required for SV40-DNA replication and allows amplification of the transfected DNA to high copy numbers, ensuring a high-level expression of the heterologous gene (e.g. levels of ~ 1 μg IL-2/10^6 cells). COS or HEK293T cells can be easily transfected by different methods such as the DEAE–dextran, the calcium phosphate, and the lipofection methods, routinely reaching transfection efficiencies from 40 to 80% (our unpublished observations). Transient expression systems may be advantageous for the synthesis of gene products that show toxicity to the host cells.

4.1.2 Permanent expression

For larger scale production purposes, the massive transfection of mammalian cells is impractical. However, it is possible to construct cell lines that constitutively secrete certain cytokines fairly efficiently (74). For example, stable cell lines were obtained secreting up to 10 μg/ml IFN-γ in the culture medium by co-transfection of CHO (Chinese hamster ovary) dhfr⁻ cells with the plasmid pAdD26SV(A)-3 (75), containing a selectable dhfr marker, and the plasmid pSV2-IFN-γ containing an expression cassette of the gene of interest, in this case huIFN-γ (76). Because the genes (selective marker and expression cassette) often become linked during the co-transfection process, there is the possibility of co-amplification of the expression cassette by selection for amplification of the *dhfr* gene during growth in increasing concentrations of methotrexate (77). In certain cases, however, additional amplification efforts did not increase the initially obtained (rather low) production levels (78). This may be the result of over-growth by low-expressing cells. Therefore, new vectors for stable expression in CHO cells were developed, producing both selectable marker and recombinant cDNA from a single primary transcript via differential splicing (79). The major drawbacks of the animal cell-culture approach are the high cost of media, the requirement of highly skilled labour, and the possibility of contamination by viruses and other cytokines.

4.2 Expression in insect cells

Insect cells, such as the widely used *Spodoptera frugiperda*-derived cell lines Sf9 and Sf21, behave very much as mammalian cells. Many of the post-translational modifications such as secretion and glycosylation are also carried out in these cells. However, mammalian cells extensively modify the core N-linked oligo-saccharide during the final glycosylation events, resulting in the formation of complex, branched oligosaccharides. In contrast, insect cells appear to be poor in this ability; in addition, their N-linked glycosyl groups have been shown to contain variable numbers of mannose residues (80). These carbohydrate struc-tures do not prevent a rapid clearing of such proteins after injection into the bloodstream (81). Insect cells have been used for the heterologous production of several cytokines after cloning the genes in a baculovirus vector (82–87). These vectors often make use of the *Autographa californica* MNPV (multiple nuclear polyhedrosis virus) very late polyhedrin or p10 promoters. In a second step the vector is recombined with the virus itself, by means of co-transfection in insect cells, to result in the production of infectious recombinant virus. High-titre virus stocks are then used to infect insect cells leading to the synthesis of the cloned gene and the death of the host. Hence, this discontinuous expression system has the disadvantage that the protein is synthesized while the cells are in the process of dying, thereby reducing the quality and the yield of the ex-pressed protein. To tackle this problem a series of vectors have been con-structed that allow continuous, high-level expression in insect cells based on the immediate–early promoter *OpIE2* (88). For example, huIL-6 was harvested from such a continuous expression system in serum-free medium at regular intervals with a cumulative yield of ~ 12 μg/ml (89).

4.3 Expression in lower eukaryotes

Because cytokines are mostly secreted proteins, their native structure is formed during a delicate and complicated translocation process through the secretion pathway. Therefore, the heterologous production of cytokines is best achieved using a host with an active secretion system. An elegant secretion system has been worked out for the yeast *Saccharomyces cerevisiae* (90). The heterologous protein is fused to the mating factor prepro-leader region (see *Figure 2*) which contains processing signals for the yeast proteases KEX2 and STE13. The KEX2 protease cleaves the precursor at the carboxyl-terminal side of Lys–Arg, and the STE13 gene product, the dipeptidyl aminopeptidase A, removes the Glu–Ala dipeptides at the amino-terminus of the excised protein in case they are present. The presence of the Glu–Ala dipeptides is optional. They will be inserted only in fusion situations where defective KEX2 cleavage is encountered.

(A) pUDT2 secretion vector

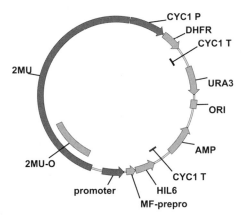

(B) Sequence of the secretion vector gene fusions

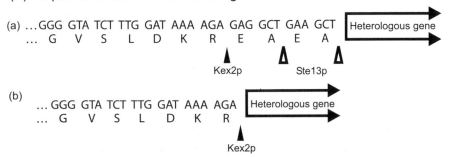

Figure 2 Structure of the *S. cerevisiae* pUDT2 secretion vector. (A) The pUDT2 vector contains: (1) the total 2-mu plasmid for replication in cir° strains; (2) the dhfr amplification cassette; (3) the ura3 selective marker; (4) the bacterial selection and replication sequences (Amp and *ori*); and (5) the secretion cassette (promoter, MF-prepro, CYCC terminator). (B) The precise structure of the fusion between the mating-factor prepro-sequence and the heterologous gene is shown. Depending on the type of fusion, the KEX2 and STE13 proteases (a) or only KEX2 enzyme (b) are involved in the processing of the fusion precursor protein.

Table 1 Expression levels obtained for cytokines using different promoters in the *S. cerevisiae* secretion system

Heterologous gene	Promoter	Yield (µg/ml)[a]
huIL-2	α-MF[b]	6.0
muIL-2	α-MF	0.1
	GAL-1	3.0
muIL-5	α-MF	0.3[c]
huIL-6	α-MF	5.0
	GAL1	30.0

[a] Biological activity measured in the non-purified medium of a yeast culture grown to an $OD_{600} = 6$.

[b] α-Mating factor.

[c] Data taken from ref. 78.

We have used the mating-factor secretion system for the expression of huIL-2 and muIL-2, muIL-5 and huIL-6 (91). The amount of cytokine secreted by the yeast cells is shown in *Table 1*. The use of the strong and well-regulated GAL1 promoter is often beneficial or even essential when expression of the heterologous gene confers counterselective pressure. Worth noting is the observation that the formation of homodimeric IL-5, which is crucial for biological activity by oxidation of two conserved cysteine residues, occurs faithfully in the yeast secretion system (78). It should also be mentioned that the glycosyl groups synthesized by the yeast cells are different from their mammalian counterparts. This may be a limitation when *in vivo* use of the recombinant protein is considered.

Another expression host, which is becoming increasingly popular, is the methylotrophic yeast *Pichia pastoris* (92). The secretion system that has been mainly used is also based on the mating-factor prepro-leader region of *S. cerevisiae*. The main differences with *S. cerevisiae* are the use of integrative vectors, e.g. pPIC9 (Invitrogen) giving the advantage of mitotic stability, and the strong, methanol-induced AOX1 promoter. Besides the superior growth and secretion capacities, *P. pastoris* also has the advantage of lower (hyper)glycosylation compared to *S. cerevisiae*.

In our laboratory, the *Pichia* secretion system has been evaluated for the expression of muIL-4, muIL-2, muIFN-β, and huIL-6; the results of which are summarized in *Table 2*. It should be noted that with the pPIC9 vector one can expect at least one extra N-terminal amino-acid residue (tyrosine) on the recombinant product, provided that the heterologous gene has been fused to the prepro-sequence using the blunt *Sna*BI restriction site. To mimic the wild-type, mature N-terminus we optimized the pPIC9 vector resulting in the plasmid pPIC92 (see *Figure 3A*). This vector contains a unique *Nae*I restriction site in which the gene of interest can be cloned so that it is directly fused to the prepro-leader without extra codons. As an example, the mature part of the muIFN-β coding sequence was ligated into the *Nae*I opened plasmid pPIC92, resulting in the

Table 2 Expression levels obtained for cytokines using the expression/secretion system of *Pichia pastoris* (strain GTS115) and some characteristics of the recombinant product

Heterologous gene (*P. pastoris* strain)	Used vector	No of copies	Yield[a] (μg/ml)	Remarks
muIL-4 (GSML1)	pPIC9	22	0.1–0.3 (active) 10 (total)	Mainly inactive possibly due to hyperglycosylation.
muIL-2 (GSML9)	pPIC9	3	60–70	Partial (±10%) STE13 processing due to *O*-glycosylation on the proximal threonine residue.
muIFN-β (GSML24)	pPIC92	ND	1.3	Homogeneous *N*-glycosylation. Correct processing: *N*-terminus is identical to the wild-type natural product.
huIL-6 (GSML20)	pPIC9	7	250	*N*-terminal processing observed by a X-prolyl dipeptidyl aminopeptidase (removal of N-terminal YP-dipeptide). A resected recombinant huIL-6 form is seen upon depletion of carbon source. See Figure 4. Both forms are fully active.

[a] Biological activity in the non-purified medium.

ND, not determined

integrative expression vector pPP1muIFN-βm which was used for the heterologous secretion of muIFN-β in *P. pastoris* (see *Figure 3B*).

In general, the characteristics of the methylotrophic yeast *Pichia pastoris* were favourable for the production of the tested recombinant cytokines. Especially huIL-6 (see *Figure 4*), and to a lower extent muIL-2, were secreted at high levels into the medium. Yields of 1–2 orders of magnitude lower were achieved with muIFN-β and muIL-4. In the case of muIL-4, it was proven that *P. pastoris* could only partially solve the glycosylation problem. In our laboratory, initial steps have been undertaken in tackling this overglycosylation problem by the co-expression of α-1,2-mannosidase from *Trichoderma reesei* (93).

Successful *P. pastoris* production systems for various cytokines have been described in the literature for huIL-17 (94), ovine IFN-τ (95), rat TNF (96), and for growth factors such as huVEGF and muEGF (97, 98). Other elegant approaches make use of a fusion of a cytokine and a soluble form of its receptor (99, 100).

4.4 Expression in bacterial systems

From the very early days of cytokine cloning, bacterial expression systems have been most helpful in providing tools for the cheap production of abundant amounts of mammalian protein. For this purpose, the mature cytokine gene (i.e. without the signal sequence information) is placed downstream of a powerful promoter ribosome-binding region. Both λP$_L$ and T7 are among the strongest promoters known in *E. coli* and allow a free choice of host strains. Versatile, tightly controlled expression vectors have been reported and used for the production of cytokines (101–103; N. Mertens *et al.*, in preparation). The pET System (Novagen) is another widely used T7 promoter-based *E. coli* expression

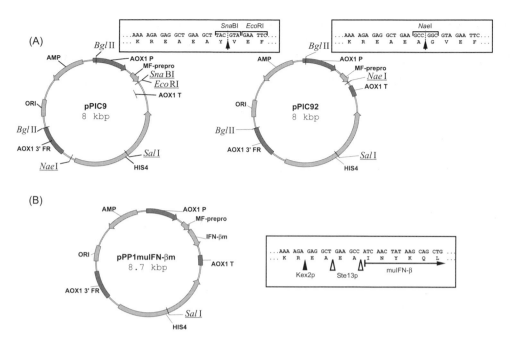

Figure 3 Structure of integrative secretion vectors for *P. pastoris*. All shown plasmids contain: the alcohol oxidase 1 promoter (AOX1 P), tightly regulated by methanol; MF-prepro, the mating-factor prepro-leader sequence of *S. cerevisiae*; AOX1 T, transcription terminator of the *AOX1* gene; HIS4, selection marker to use in his4 auxotrophic *P. pastoris* strains; AOX1 3′ FR, segment downstream of the *AOX1* gene which allows *AOX1* gene replacement; AMP, β-lactamase and ORI, replication origin for bacterial selection and replication. (A) In pPIC9 (Invitrogen) the blunt *Sna*BI site was used for cloning purposes, resulting in an extra tyrosine residue upon processing by the KEX2 and STE13 proteases. We optimized pPIC9, resulting in pPIC92. In the latter, the heterologous gene can be cloned into the blunt *Nae*I site so that precise fusion is possible, without the extra amino-terminal tyrosine residue. In (B) the result is shown in the case of muIFN-β.

system (104). More than 40% of the total protein in the culture may correspond to the protein of interest. A complication that often occurs, however, is that this heterologous protein is present in the bacterial cell as insoluble inclusion bodies. Procedures must then be followed to dissolve these in chaotropic media, followed by renaturation and extensive purification (see *Figure 5*). One of the exceptions is the bacterial expression of huTNF from the heat-inducible P_L promoter, where a production of 44 μg/ml of soluble, fully active, homotrimeric protein was obtained when bacteria were grown in shake flasks (14). Various mammalian cytokines have been produced in *E. coli* as soluble thioredoxin fusion proteins (105). Several of these cytokines had previously been produced as insoluble inclusion bodies.

To prevent the problem of insoluble inclusion bodies one may think of using expression vectors that direct the secretion of the recombinant cytokine into the periplasmic space, from which it can easily be recovered and purified. This can be obtained by constructing fusion proteins between the cytokine and the

(A)

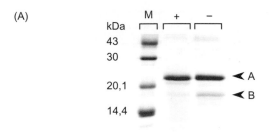

(B)

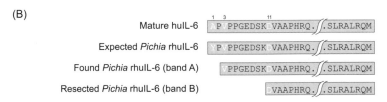

Figure 4 Expression of huIL-6 in *P. pastoris*. (A) Comparison of proteolytic degradation of the recombinant huIL-6 production strain GSML20 after 48 h of induction, with (+) or without (–) the addition (every 12 h) of extra methanol as inducer and carbon source. 20 μl of crude medium was loaded on to a PAGE gel. The amino-terminal sequences of the two obtained huIL-6 bands A and B were determined (B). Both recombinant huIL-6 forms had other N-termini than expected. The recombinant huIL-6 was trimmed by an X-prolyl dipeptidyl aminopeptidase (X-prolyl DPAP, EC 3.4.14.5) which removed the YP-dipeptide, producing band A. The next dipeptide is not removed because the third residue is proline (the recognition sequence of the enzyme is XxxPro®Yyy with Yyy ≠Pro or hydroxyPro). In our laboratory it was shown that the gene product of STE13 is responsible for the hydrolysis of the AlaPro N-terminal dipeptide of recombinant huIL-6 produced in *S. cerevisiae* (91). Therefore it seems most likely that this X-prolyl DPAP activity comes from the *P. pastoris* STE13. The huIL-6 B-form is the product of a proteolytic enzyme which is released into the medium, probably upon cell lysis due to depletion of the carbon source. Both forms are fully active and can not be resolved with standard protein purification techniques (unpublished results).

secretion signal sequence of *E. coli* OMP A reaching expression levels of up to 4 μg/ml per OD_{600} (84, 106). In addition, bacterial strains have been developed that enable disulfide bond formation in the *E. coli* cytoplasm (107; Novagen).

As a general rule, it is our belief that overloading the protein production machinery of *E. coli* by trying to express huge quantities of cytokines is often disadvantageous to the system, and that this results in insoluble recombinant protein. More moderate production speed and/or levels may contribute to the solubility of the expressed protein.

5 Conclusions

Cytokine genes have been isolated using a wide range of elegant cloning techniques. Due to the improvements in mRNA preparation methods, cDNA synthesis technology, and transfection procedures, the direct expression methods become more and more attractive for the cloning of new biological activities. In modern

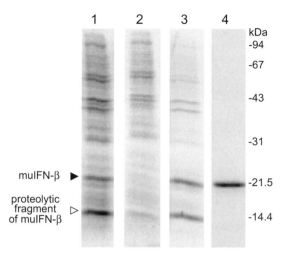

muIFN-β ▶

proteolytic
fragment ▷
of muIFN-β

Figure 5 Bacterial expression of muIFN-β. SDS-PAGE of *E. coli* expressed muIFN-β showing complete bacterial lysate (french press, lane 1), soluble (lane 2), and insoluble (lane 3) fractions of *E. coli* proteins and purified muIFN-β (lane 4). Note that the expressed IFN-β is contained in the insoluble fraction (inclusion bodies). Together with the expected, full-length muIFN-β (filled arrow) a proteolytic breakdown product of muIFN-β (open arrow) is present in the inclusion bodies. The insoluble fraction was dissolved in 7 M guanidinium thiocyanate, renatured, and muIFN-β purified to homogeneity. The purified muIFN-β has a specific biological activity of approximately 6×10^7 U/mg in an antiviral assay.

biology, the WWW (world wide web) gives easy access to data generated by major sequencing efforts such as the EST sequencing and the genome projects. This *in silico* methodology allows the screening and identification of novel cDNAs related to known cytokines.

Abundant synthesis of protein in heterologous systems can be obtained for all cytokines. Often this is a scenario of trial and error. Depending on the purpose for cloning the gene for *in vivo* or *in vitro* experiments, and the requirement for secondary modification, one can choose from the more reliable, but more expensive mammalian or insect expression systems, yeast secretion systems, or bacterial expression.

Acknowledgements

The VIB, IUAP, and the Belgian Fund for Medical Research (FGWO) supported research in the authors' laboratory. The authors are grateful to S. Dewaele for contribution of protocols, F. Duerinck for muIFN-β data, and W. Drijvers for art work.

References

1. Altschul, S. *et al.* (1990). *J. Mol. Biol.*, **215**, 403.
2. Altschul, S. F. *et al.* (1997). *Nucleic Acids Res.*, **25**, 3389.

3. Sambrook, J., Fritsch, E. F., and Maniatis, T. (1989). *Molecular cloning. A laboratory manual.* Cold Spring Harbor Laboratory Press, New York.

4. Janssen, K. (1994). *Current protocols in molecular biology.* Wiley, Boston, MA.

5. Ausubel, F. M. *et al.* (ed.) (1990). *Current protocols in molecular biology.* Green Publishing Associates and Wiley-Intersience, New York.

6. Contreras, R. *et al.* (1995). In *Cytokines: a practical approach* (ed. D. Rickwood and B. D. Hames). IRL Press, Oxford.

7. Kurybayashi, K. *et al.* (1988). *Nucleic Acids Res.,* Symposium Series, **19**, 61.

8. Hornes, E. and Korsner, L. (1990). *Genet. Anyl. Tech. Appl.*, **7**, 145.

9. Nagata, S. *et al.* (1980). *Nature,* **287**, 401.

10. Devos, R. *et al.* (1983). *Nucleic Acids Res.,* **11**, 4307.

11. Taniguchi, T. *et al.* (1983). *Nature,* **302**, 305.

12. Pennica, D. *et al.* (1984). *Nature,* **312,** 724.

13. Fransen, L. *et al.* (1985). *Nucleic Acids Res.,* **13**, 4417.

14. Marmenout, A. *et al.* (1985). *Eur. J. Biochem.,* **152,** 515.

15. Nagata, S. *et al.* (1986). *Nature,* **319,** 415.

16. Miyamoto, M. *et al.* (1993). *Mol. Cell. Biol.,* **13**, 4251.

17. Bartley, T. *et al.* (1994). *Cell,* **77**, 1117.

18. Okamura, H. *et al.* (1995). *Infect. Immun.,* **63**, 3966.

19. Contreras, R. *et al.* (1982). *Nucleic Acids Res,* **10**, 6353.

20. Noma, Y. *et al.* (1986). *Nature,* **319,** 640.

21. Kinashi, T. *et al.* (1986). *Nature,* **324**, 70.

22. Yokota, T. *et al.* (1984). *Proc. Natl. Acad. Sci. USA,* **81,** 1070.

23. Lee, F. *et al.* (1985). *Proc. Natl. Acad. Sci. USA,* **82,** 4360.

24. Goodwin, R. G. *et al.* (1993). *Eur. J. Immunol.,* **23**, 2631.

25. Minty, A. *et al.* (1993). *Nature,* **362**, 248.

26. Heinrich, J. N. *et al.* (1993). *Mol. Cell. Biol.,* **13**, 2020.

27. Rouvier, E. *et al.* (1993). *J. Immunol.,* **150**, 5445.

28. Wong, B. *et al.* (1997). *J. Biol. Chem.,* **272**, 25190.

29. Tashiro, K. *et al.* (1999). In *Methods in enzymology,* Vol. 303 (ed. M. Weissman) p. 479. Academic Press, London.

30. Imai, T. *et al.* (1996). *J. Biol. Chem.,* **271**, 21514.

31. Lomedico, M. *et al.* (1984). *Nature,* **312**, 458.

32. Noma, Y. *et al.* (1986). *Nature,* **319,** 640.

33. Kinashi, T. *et al.* (1986). *Nature,* **324**, 70.

34. Kothakota, S. *et al.* (1997). *Science,* **278**, 294.

35. Mierendorf, R. *et al.* (1987). In *Methods in enzymology,* Vol. 152, (ed. S. Berger and A. Kimmel), p. 458. Academic Press, London.

36. Ambrus, J. L. *et al.* (1993). *Proc. Natl. Acad. Sci. USA,* **90**, 6330.

37. Arrufo, A. *et al.* (1990). *Cell,* **61**, 1303.

38. Loetscher, H. *et al.* (1991). *J. Biol. Chem.,* **266**, 18324.

39. Suda, T. *et al.* (1993). *Cell,* **75**, 1169.

40. Hannum, C. *et al.* (1994). *Nature,* **368**, 643.

41. Okayama, H. and Berg, P. (1982). *Mol. Cell. Biol.,* **2,** 161.

42. Gheysen, D. and Fiers, W. (1982). *J. Mol. Appl. Genet.,* **1,** 385.

43. Huylebroeck, D. *et al.* (1988). *Gene,* **66**, 163.

44. Seed, B. and Arrufo, A. (1987). *Proc. Natl. Acad. Sci. USA,* **84**, 3365.

45. Miyazaki, J. *et al.* (1989). *Gene,* **79**, 269.

46. Wong G. *et al.* (1985). *Science,* **228,** 810.

47. Yang Y.-C. *et al.* (1986). *Cell,* **47,** 3.

48. Nagasawa, T. *et al.* (1994). *Proc. Natl. Acad. Sci. USA*, **91**, 2305.
49. Lok, S. *et al.* (1994). *Nature*, **369**, 565.
50. Cruikshank, W. *et al.* (1994). *Proc. Natl. Acad. Sci. USA*, **91**, 5109.
51. Sundick, R. and Gill-Dixon, C. (1997). *J. Immunol.*, **159**, 720.
52. Aggarwal, B. B. and Natarajan, K. (1996). *Eur. Cytokine Netw.*, **7**, 93.
53. Raport, C. J. *et al.* (1996). *J. Leucocyte Biol.*, **59**, 18.
54. Power, C. *et al.* (1997). *J. Exp. Med.*, **186**, 825.
55. Lyman, S. *et al.* (1993). *Cell*, **75**, 1157.
56. Rayner, J. R. and Gonda, T. J. (1994). *Mol. Cell. Biol.*, **14**, 880.
57. Wong, B. *et al.* (1994). *J. Virol.*, **68**, 5523.
58. Lathe, R. (1985). *J. Mol. Biol.,* **183,** 1.
59. Martin, F. H. and Castro, M. M. (1985). *Nucleic Acids Res.,* **13**, 8927.
60. Wallace, R. B. and Miyada, C. G. (1987). In *Methods in enzymology*, Vol. 152 (ed. S. Berger and A. Kimmel), p. 432. Academic Press, London.
61. Kawasaki, E. *et al.* (1985). *Science,* **230,** 291.
62. Gough, N. *et al.* (1984). *Nature,* **309,** 763.
63. Hirano, T. *et al.* (1986). *Nature,* **324,** 73.
64. Grabstein, K. *et al.* (1994). *Science*, **264**, 965.
65. Okamura, H. *et al.* (1995). *Nature*, **378**, 88.
66. Loetscher, H. *et al.* (1990). *Cell*, **61**, 351.
67. Samal, B. *et al.* (1994). *Mol. Cell. Biol.*, **14**, 1431.
68. Hilton, D. *et al.* (1994). *EMBO*, **13**, 4765.
69. Yao, Z. *et al.* (1995). *J. Immunol.*, **155**, 5483.
70. Power, C. *et al.* (1996). *Trends Pharmacol. Sci.*, **17**, 209.
71. Van de Craen, M. *et al.* (1997). *FEBS Lett.*, **403**, 61.
72. Ware, C. *et al.* (1998). In *The cytokine handbook* (ed. A. Thomson), p. 549. Academic Press, London.
73. Bacon, K. *et al.* (1998). In *The cytokine handbook* (ed. A. Thomson), p. 753. Academic Press, London.
74. Levinson, A. and (1991). In *Methods in enzymology*, Vol. 185 (ed. D. Goeddel), p. 485. Academic Press, London.
75. Kaufman, R. and Sharp, P. (1982). *Mol. Cell. Biol.,* **2,** 1304.
76. Scahill, S. *et al.* (1983). *Proc. Natl. Acad. Sci. USA,* **80,** 4654.
77. Cullen, B. R. (1987). In *Methods in enzymology*, Vol. 152 (ed. S. L. Berger and A. R. Kimmel)., p. 684. Academic Press, London.
78. Tavernier, J. *et al.* (1989). *DNA,* **8,** 491.
79. Lucas, B. *et al.* (1996). *Nucleic Acids Res.*, **24**, 1774.
80. Kuroda, K. T. *et al.* (1990). *Virology*, **174**, 418.
81. Sareneva, *et al.* (1993). *J. Interferon Res.*, **13**, 267.
82. Morelle, C. *et al.* (1993). *Biofutur*, **125**, 1.
83. Guisez, Y. *et al.* (1995). *FEBS Lett.*, **331**, 49.
84. Guisez, Y. *et al.* (1998). *Protein Expr. Purif.*, **12**, 249.
85. Lambrecht, B. *et al.* (1999). *Vet. Immunopathol.*, **70**, 257.
86. Hui, S. *et al.* (1998). *Chin. J. Biotechnol.*, **14**, 205.
87. James, D. *et al.* (1996). *Protein Sci.*, **5**, 331.
88. Pfeifer, T. (1998). *Curr. Opin. Biotechnol.*, **9**, 518.
89. Invitrogen (1999). In *Expressions, a newsletter for gene cloning and expression*, Vol. 6, No. 4, p. 6.
90. Brake, A. *et al.* (1984). *Proc. Natl. Acad. Sci. USA*, **81**, 4642.

91. Guisez, Y. *et al.* (1991). *Eur. J. Biochem.,* **198**, 217.
92. Hollenberg, C. and Gellissen, G. (1997). *Curr. Opin. Biotechnol.*, **8**, 554.
93. Martinet, W. *et al.* (1998). *Biotechnol. Lett.*, **20**, 1171.
94. Murphy, K. *et al.* (1998). *Protein Expr. Purif.*, **12**, 208.
95. Van Heeke, G. *et al.* (1996). *J. Interferon Cytokine Res.*, **16**, 119.
96. Rees, G. *et al.* (1999). *Eur. Cytokine Netw.*, **10**,383.
97. Mohanraj, D. *et al.* (1995). *Growth Factors*, **12**, 17.
98. Clare, J. *et al.* (1991). *Gene*, **105**, 205.
99. Fischer, M. *et al.* (1997). *Nature Biotechnol.*, **15**, 142.
100. Pflanz, S. *et al.* (1999). *FEBS Lett.,* **450,** 117. [Published erratum appears in (1999). *FEBS Lett.*, **454**, 172.]
101. Mertens, N. *et al.* (1995). *Gene*, **164**, 9.
102. Mertens, N. *et al.* (1995). *Bio/Technol.*, **13**, 175.
103. Mertens, N. *et al.* (1999). In *Biotechnology international II: latest developments in the biotechnology industry and research*, Vol. 2 (ed. T. Connor, H. Weier, and F. Fox), p. 165. Universal Medical Press, San Francisco, CA.
104. Studier, F. *et al.* (1990). In *Methods in enzymology,* Vol. 185 (ed. D. Goeddel), p. 60. Academic Press, London.
105. LaVallie, E. *et al.* (1993). *J. Biol. Chem.*, **268**, 23311.
106. De Sutter, K. *et al.* (1994). *Gene*, **141**, 163.
107. Derman, A. *et al.* (1993). *Science*, **262**, 1744.

Chapter 2

Detection and population analysis of IL-1 and TNF gene polymorphisms

Francesco S. di Giovine, Nicola J. Camp, Angela Cox, Adeel G. Chaudhary, Julian A. Sorrell, Alison Crane, and Gordon W. Duff

University of Sheffield, Division of Molecular and Genetic Medicine, M Floor, Royal Hallamshire Hospital, Glossop Road, S10 3GZ, Sheffield, U.K.

1 Introduction

The pro-inflammatory cytokines IL-1 and TNF-α mediate inflammatory responses by attracting, and activating, white blood cells to tissues, and stimulating the secretion of other lymphocytotropic cytokines and catabolic enzymes (1, 2). In acute infection the inflammatory response helps to contain and eradicate the pathogen; in chronic inflammatory diseases, production of these cytokines causes tissue damage.

The question of what stimulates the inappropriate production of these cytokines in autoimmune or infectious diseases awaits a clearer understanding of the pathogenic mechanisms of disease. Whatever the environmental factors may be, it is clear that in most chronic inflammatory diseases there is a genetic component of host response which cannot entirely be attributed to the immune response genes within the Major Histocompatibility Complex (MHC) (3).

1.1 Genetic variation at IL-1 and TNF loci

If IL-1 and TNF are centrally involved in inflammation, they may be important in inflammatory diseases irrespective of the initiating stimulus and the anatomical site of the cellular reaction. In recent years, we have acted on the premise that IL-1 and TNF are important mediators of inflammation and have asked the question: Is genetic variation in these cytokine genes relevant to clinical inflammatory diseases? To tackle this we first searched for DNA variants of IL-1 and TNF by sequencing, then performed case-control association studies to test allele frequencies in disease populations compared with ethnically matched control

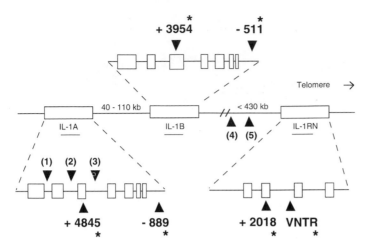

Figure 1 Structure of the interleukin-1 locus. The IL-1 locus is within the long arm of chromosome 2 (2q13). IL-1A, IL-1B, and IL-1RN all lie within a 430 kb region. Polymorphic sites treated in detail in this chapter are indicated by their position in the genes and an asterisk, other polymorphisms are numbered and some details can be found in *Table 1*.

Table 1 Nature of gene variant

		Ref.
1	VNTR, intron 6 (86 bp repeats)	64
2	Dinucleotide repeat, intron 5	65
3	Trinucleotide repeat, intron 4	66
4	Tetranucleotide repeat	67
5	Dinucleotide repeat	68

populations. There are three known genes in the IL-1 gene family (*IL1A*, *IL1B*, and *IL1RN*), situated in a 430-kb region on the long arm of human chromosome 2 (4) (*Figure 1*, *Table 1*). The *IL1A* and *IL1B* gene products, IL-1α and IL-1β, bind to the IL-1 receptor type I (5) and initiate signalling by a pathway that seems homologous with the *Toll* pathway in *Drosophila* spp. (6). The product of *IL1RN*, i.e. IL-1 receptor antagonist (IL-1ra) (7, 8), also binds the receptor but does not initiate intracellular signalling, so acting as a competitive inhibitor (9). *In vitro* and *in vivo* IL-1ra has powerful anti-inflammatory effects (10). The gene coding for TNF-α (*TNFA*) is on human chromosome 6 in the class III region of the MHC in tandem with lymphotoxin A and B (*Figure 2*, *Table 2*).

It is beyond the scope of this chapter to describe detailed protocols for the detection of novel gene variants. Diverse techniques can be used, including chemical cleavage (11), MutS detection (12), denaturing gradient gel electrophoresis (DGGE) (13), and heteroduplex analysis (14). In our hands, single-stranded conformation polymorphism (SSCP) (15) and direct sequencing have been the most productive in the past.

We have described several polymorphisms in the three IL-1 genes (*IL-1A*, *IL-1B*, *IL-1RN*) and in *TNFA*. In this chapter we first summarize our studies on the

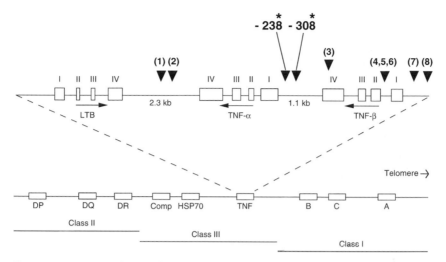

Figure 2 Structure of the human Major Histocompatibility Complex. The MHC locus spans 4–5 Mb on the short arm of chromosome 6 (6q21). It houses the genes for other families of proteins intimately associated with immunological recognition and activation, including the TNF family. The polymorphic sites treated in detail in this chapter are indicated by their position in respect to TNF transcription initiation site and an asterisk, other sites relevant to the TNF locus are numbered consecutively. Details of these are in the accompanying *Table 2*.

Table 1 Nature of gene variant

		Ref.
1	Dinucleotide repeat	59
2	Dinucleotide repeat	59
3	Single-base variation	60
4	Dinucleotide repeat	61
5	Single-base variation	62
6	Single-base variation	63
7	Dinucleotide repeat	61
8	Dinucleotide repeat	61

frequency of these normal gene variants in case-control studies of human inflammatory diseases, and then describe routine screening methods for these and other gene variations in the IL-1 and TNF cluster for the purpose of genetic association and family studies. In the last part of this chapter a few notes on how to avoid frequent mistakes can be found, including some statistical advice on genetic association studies.

2 Case-control studies for IL-1 and TNF gene polymorphism

Alleles of the IL-1 genes are associated with several autoimmune and inflammatory diseases, where they may have a role in determining the severity of the disease as well as possibly affecting susceptibility to disease. Allele 2 of the variable number of tandem repeats (VNTR) in the IL-1 receptor antagonist gene (*IL1RN*) was the first marker of the IL-1 cluster to be associated with severity in systemic lupus erythematosus (SLE), ulcerative colitis, alopecia areata, and lichen sclerosus (16–19). In SLE, for example, whilst there is an overall association with allele 2, the association can probably be accounted for by those patients who fulfil more than six American Rheumatism Association (ARA) criteria, and in those with photo-sensitivity and discoid lesions (16). In ulcerative colitis, the association is strongest in those with total colon involvement (17); in alopecia, in those with alopecia universalis (18), and in lichen sclerosus in those with extragenital lesions (19). It seems plausible, then, that the *IL1RN* allele 2 may be a marker for the severity of some inflammatory diseases.

Since the VNTR is in an intron of the gene, it is possible that different VNTR alleles have a direct effect on the expression of the *IL1RN* gene itself, or the VNTR may be acting as a marker for another linked functional polymorphism. With regard to direct functional effects, data thus far are conflicting. Two groups have identified allele-specific differences in gene expression with various stimuli, although in opposite directions (20, 21). In contrast, using allele-specific tran-script analysis with a linked exonic marker, Clay *et al.* were unable to detect any allele-specific differences in mRNA accumulation in heterozygous keratinocyte cell lines (22). It therefore seems unlikely that the VNTR is having a significant effect on the production of IL-1 receptor antagonist protein, and thus it is possible that allele 2 is acting as a marker for a nearby gene. This is consistent with the observation that in ulcerative colitis, for example, the VNTR association is seen in some population studies and not in others, perhaps reflecting differing patterns of linkage disequilibrium in different ethnic groups (23).

The other two genes of the IL-1 cluster have also been associated with disease. A biallelic marker at position –889 of the IL-1 alpha gene was associated with juvenile chronic arthritis (early-onset, pauciarticular patients), and allele 2 of this polymorphism was also associated with an inflammatory complication of the disease, higher allele frequencies being observed in those individuals with chronic iridocyclitis (relative risk = 6.2) and those with higher erythrocyte sedi-mentation rate (ESR) measurements in the first year of clinic attendance (24). The −889 polymorphism is in linkage disequilibrium with a coding region poly-morphism at +4845 in the IL-1 alpha gene, and it remains to be determined whether either of these affect the function of the gene in terms of altered protein or increased protein production. In the IL-1 beta gene, however, there is evidence of a relationship between a polymorphism and rate of protein production. A *Taq* I restriction fragment length polymorphism (RFLP), now known to be in intron V of the IL-1 beta gene, is associated with insulin-dependent diabetes mellitus (IDDM) in HLA DR3/4-negative patients, and the same polymorphism has been correlated with IL-1 beta protein production from bacterial lipopolysaccharide (LPS)-stimulated monocytes (25). In addition, allele 2 of this marker is associated in a gene dose-dependent fashion with psoriasis (26) and periodontal disease (27), consistent with the association with protein production (26). Markers in the IL-1 cluster have also been associated with psoriasis (26), idiopathic pulmonary fibrosis (28), erosive rheumatoid arthritis (29), and a lethal outcome in meningitis (30).

Given the complexity of the IL-1 system, it will take some time to elucidate the exact relationships between the different markers and the inflammatory phenotype, which will involve the IL-1 receptors in addition to the IL-1 genes themselves. A high degree of linkage does exist in this region (31) and it is likely that gene functional studies will be complex to perform and extremely difficult to assess. Nevertheless, it is clear that the IL-1 family of genes has an important role to play in the clinical outcome of inflammatory diseases and that genetic studies in this cluster can be useful in the assessment of risk and management of patients.

The *TNF* locus in the class III region of the MHC is also a good candidate gene in autoimmune and inflammatory diseases, but because of the high degree of linkage disequilibrium across the MHC, it is difficult to determine which genes on a haplotype are important in the aetiology of a disease. The haplotype HLA-A1-B8-DR3-DQ2, known as the 'autoimmune haplotype' is associated with a number of autoimmune diseases, including insulin-dependent diabetes, Graves' disease, myasthenia gravis, SLE, dermatitis herpetiformis, and coeliac disease (32–34). A biallelic polymorphism at position –308 of the TNF alpha promoter has been studied in these diseases, since it has been shown that: (a) high TNF alpha production levels have been associated with particular DR3 and DR4 haplotypes (35); and (b) that the TNF2 allele at –308 is carried on the autoimmune haplotype (36). However, in all the diseases mentioned above, it has not been possible to demonstrate any association of TNF with disease independently of the association with the autoimmune haplotype.

It seems that TNF does have an important role to play in infectious diseases; in a large study of patients with malaria in the Gambia, TNF2 homozygosity was strongly associated with death from cerebral malaria, and no association with clinical outcome was found with any other marker in the class I and II regions of the MHC (37). Investigations of other infectious diseases will be very interesting in this regard.

The results from population-based association studies with candidate genes are useful for confirming candidate gene status, and as a starting point for functional studies. Some advice on the set-up and analysis of association studies can be found in the last part of this chapter. However, additional data is required to confirm linkage of the candidate gene region, involving family-based studies to demonstrate segregation of the gene of interest with the disease. Once linkage is confirmed, linkage-disequilibrium mapping can be carried out to fine-map the region of maximum association with the disease, for example using a panel of microsatellite markers spanning the region of interest for transmission-disequilibrium testing (see below) (38).

3 Methods for allelic discrimination in the IL-1 and TNF gene clusters

Prior to sample collection, informed consent is needed from all subjects. It is recommended that the local Ethics Committee is consulted, in order to advise on the correct procedures and how to handle genetic data in accordance with legal requirements such as the Data Protection Act in the UK.

Genotyping can have forensic and diagnostic relevance in addition to scientific interest, so that appropriate quality controls, validation experiments, and caution must be applied at all stages (see Section 4.1 and following).

3.1 Preparation of template DNA

Polymerase chain reaction (PCR)-based genotyping does not require particularly high-MW DNA (< 20 kb DNA is often an excellent template). As 100 ng of genomic DNA is more than sufficient for single-copy gene amplification, direct amplification from dried blood spots or cell lysates can be used for genotyping (39), and two of the protocols that we have used are described below (*Protocols 1* and *2*).

However, if DNA banks need to be established for population studies where DNA needs to be stored for future reference or genotyping at different loci, or where genomic Southern blotting might be needed, good quality high-MW genomic DNA needs to be extracted. For this purpose *Protocol 3* (see below) has proved very reliable in our laboratory. Basic buffers and the composition of chemical solutions can also be found in major protocol textbooks (40, 41).

Protocol 1

Preparation of PCR-grade DNA from dried blood spots

NB: It is very important that each step is performed in a clean area (physically isolated from post-PCR events).

Equipment and reagents

- Clean sheets of filter paper (We have used samples from several suppliers with good results.)
- Disposable glass Pasteur pipette
- Mineral oil
- PCR mastermix (see *Protocol 4*)
- PCR tubes (0.5 or 0.2 ml thin-walled polypropylene tubes)
- Disposable scalpel blades
- Sterile needles

Method

1   Spot uncoagulated blood evenly on to a clean sheet of filter paper using a sterile Pasteur pipette.

2   Leave filters to dry overnight, store subsequently at room temperature.

3   Use a disposable scalpel blade to cut out approximately 1 mm^2 of the blood spot and place it at the bottom of each tube. Use control paper from areas of the filters which had not been spotted with blood as negative controls.

4   Prepare a mastermix as described in *Protocol 4*, but omit the addition of *Taq* polymerase. Typically, 24 μl are used for a final 25 μl PCR reaction.

5   Overlay the samples with 40 μl of mineral oil. Pierce the lid of each tube with a sterile needle.

Protocol 1 continued

6   Heat the tubes at 98 °C for 15 minutes, and then leave to cool for a few minutes.

7   Store the samples at this stage, or add 1 µl (0.25 U) of *Taq* polymerase to start PCR directly.

Protocol 2
Preparation of PCR-grade DNA from cell lysates

Equipment and reagents

- Proteinase K (PK) buffer: 0.1 M NaCl, 10 mM Tris–HCl, pH 8.0 25 mM EDTA, 0.5% SDS pH 8.0, 0.1 mg/ml fresh Proteinase K
- Polypropylene micro tubes (0.5–0.2 ml)
- Dry block
- Microcentrifuge
- Sterile needles

Method

1   Resuspend cells from saliva or white blood cells separations, in 0.2 ml of PK buffer, in 0.5 ml polypropylene tubes. Tissue samples can also be extracted, by homogenising with a Dounce or hand-held rotary homogeniser (Powergen 35, Fisher Scientific). White blood cells can easily be obtained by density centrifugation on Fycoll Hypaque (Nycomed, Sweden).

2   Pierce the tube lids with a sterile needle.

3   Incubate samples in a dry-block at 95 °C for 10 min.

4   Centrifuge the tubes at 8000 RCF in a microcentrifuge, and store the supernatants at −20 °C prior to PCR.

5   For higher quality DNA, perform a phenol/chloroform extraction followed by ethanol precipitation, according to standard protocols (40, 41).

Protocol 3
Preparation of genomic DNA from blood cells

Equipment and reagents

- Venepuncture equipment
- Sterile 15 and 50 ml polypropylene tubes
- 32 mM Tris, 10 mM Sucrose 5 mM MgCl2. Adjust pH to 8.0 with 37% HCl. After autoclaving add 1% Triton X-100
- White blood cell (WBC) lysis solution: 0.4 M Tris, 60 mM EDTA, 0.15 M NaCl, 10% SDS. Adjust pH to 8.0.
- Rotary mixer
- High-speed, high-capacity centrifuge (able to spin 6 to 24 tubes 50 ml tubes at 1300 g) e.g. Beckman J-6B with 8 × bucket rotor
- Microcentrifuge
- 'Nucleon' Silica Suspension (ScotLab, UK)
- Tris–EDTA buffer pH 7.0

Method

1 Collect blood by venepuncture and store uncoagulated at −20 °C prior to DNA extraction. When possible, we prefer to collect two 10 ml samples, extract DNA form the first sample and keep the second for future reference.

2 Transfer 10 ml of blood to 50 ml sterile polypropylene tubes, add 40 ml of the RBC lysis solution, and mix by inversion for 4 min at room temperature.

3 Centrifuge the samples at 1300 g for 15 min 4 °C, aspirate the supernatant and discard.

4 Add another 30 ml of RBC lysis solution to the cell pellet, and mix gently. Centrifuge the samples at 1300 g for 15 min 4 °C.

5 Resuspend the resulting pellet in 2 ml of the white blood cell (WBC) lysis solution and transfer into a fresh 15 ml polypropylene tube.

6 Add sodium perchlorate to a final concentration of 1 M.

7 Invert the tubes on a rotary mixer for 15 min at room temperature, then incubate at 65 °C for 25 min, invert periodically.

8 Add 2 ml of chloroform (stored at −20 °C), mix the samples for 10 min at room temperature and then centrifuge at 800 g for 3 min.

9 To obtain a very clear distinction of phases add 300 µl of the Nucleon Silica suspension and centrifuge at 1400 g for 5 min.

10 Aspirate the resulting aqueous upper layer, and transfer to a fresh 15 ml polypropylene tube.

11 Add cold ethanol (stored at −20 °C) to precipitate the DNA. (Genomic DNA will appear as a filamentous, translucent, and silvery phase in the tube.)

12 Spool the DNA out on a glass hook and transfer to a 1.5 ml Eppendorf tube, containing 500 µl TE buffer or sterile water. Leave overnight R.T. to resuspend.

13 Calculate the genomic DNA yield by spectrophotometry at 260 nm, where an optical reading of 1 OD_{260} is equal to a concentration of 50 µg/ml of double-stranded DNA.

14 Dilute aliquots of samples at 100 µg/ml, transfer to microtitre containers and store as needed at 4 °C. Store stocks at −20 °C for long-term future reference.

3.2 Polymerase chain reaction (PCR)

All the genotyping methods that we currently use are based on a polymerase chain reaction step. Oligonucleotide primers designed to amplify the relevant region of the gene spanning the polymorphic site (as detailed below) are synthesized, resuspended in Tris–EDTA buffer (TE), and stored at −20 °C as stock solutions of 200 µM. Aliquots of working solutions (1:1 mixture of forward and reverse, 20 µM of each in water) are prepared in advance of the experiment.

A typical PCR protocol is detailed below. Differences from this standard protocol will be highlighted in the individual genotyping protocols later in this chapter (see Sections 3.4.1 to 3.4.8).

Protocol 4

Polymerase chain reaction (PCR)

Equipment and reagents

- Dedicated micropipette for pre-PCR operation (for larger scale work use electronic multichannel pipettes)
- 96-well Multiwell plates, PCR-grade Advanced Biotechnologies
- Sterile water, individually packed
- TE buffer, pH 8.0
- *Taq* polymerase, licensed for use in PCR reactions (Roche)

- 10 × PCR buffer: 200 mM Tris–HCl pH 8.4, 500 mM KCl
- 50 mM $MgCl_2$ stock solution
- dNTP stock solution: 10 mM each dATP, dCTP, dGTP, dTTP in water solution
- Oligonucleotide primer solution: 20 μM in TE buffer
- DNA thermocycler (PE.ABI-MJR)

Method

1 Dot DNA template on the bottom of microwell (capacity of wells) well strips or microplates.

2 Prepare a 'mastermix' with all reagents, taking into account the number of samples, and including positive and negative controls. Typically, for a 50 μl reaction mix, use the following final concentrations: $MgCl_2$ at 1.75 mM, 0.2 mM of each dNTP, 1 μM of each PCR primer, 1.25 U *Taq* polymerase, and 100 ng of the DNA template.

3 Dispense 50 μl of the mastermix into each well of the multiwell plate, using a micropipette or multichannel pipette as required.

4 If no heated lid option is available for the DNA thermocycler, add 20 μl mineral oil to each well.

5 Cycle as required by the individual protocols, genotype-specific.

6 Store samples at fridge temperature or process as required.

3.3 PCR-RFLP genotyping

The presence of DNA polymorphisms can complete or disrupt an enzyme recognition site. Consequently, allelic forms of DNA can be discriminated because of their different restriction pattern. The use of PCR to provide limitless copies of the DNA region of interest has made PCR-RFLP methods the most used genotyping strategy for medium-to-low volume studies.

The following is a general protocol for PCR-RFLP analysis of the several IL-1 and TNFA gene polymorphisms. Specific information on individual loci (PCR primer sequences, gene accession numbers, PCR cycling conditions, etc.) are listed below.

Fragment sizing is generally performed by polyacrylamide gel electrophoresis (PAGE), but a 2% agarose horizontal gel can be used for IL-1RN (VNTR)

Protocol 5

PCR-RFLP genotyping

Equipment and reagents

- Dry-block heater
- Polypropylene microtubes (96-well) or microwell plates
- Appropriate restriction enzyme and panel buffer (Promega, New Engl. Biolabs)
- Double-distilled water
- Mineral oil (Sigma)
- Horizontal agarose gel electrophoresis apparatus (Biorad, Hoefer)
- Vertical polyacrylamide gel (Biorad, Hoefer/ Amersham) electrophoresis apparatus

- Agarose (BDH)
- Polyacrylamide/bis-acrylamide solution (BDH)
- Ethidium bromide solution. Stock solution prepared as 10 mg/ml in water. Used for staining at 0.5 μg/ml
- Tris–borate buffer TBE = 90 mM Tris-Borate 2 mM EDTA, bring pH to 8.0
- Gel documentation apparatus (Polaroid or videocamera with thermal printer)
- Gel-loading dye: 0.25% Bromophenol Blue 0.25% xylene cyanol FF 40% sucrose in water

Method

1 Transfer 25 μl of PCR products into each microtube.

2 Prepare a digestion mastermix. For each sample use 3 μl of appropriate restriction buffer, 5–10 U of enzyme required, and ddH$_2$O to give a total volume of 5 μl per sample. Typically, we prepare 10% more mastermix than required—e.g. for 20 samples we would prepare 110 μl of digestion mix, and pipette 5 μl into each microtube.

3 Dispense 5 μl of the digestion mix per sample—bring to a total volume of 30 μl.

4 Overlay with 20–40 μl of mineral oil and incubate on a dry block at the appropriate temperature overnight.

5 Aspirate 15 μl of sample and mix with 5 μl gel-loading dye.

6 Carry out polyacrylamide-gel electrophoresis (PAGE) on 20–40 μl PCR samples in Tris–Borate–EDTA buffer at constant voltage. Depending on the size discrimination needed, use different PAGE conditions (9–12% acrylamide, 1.5 mm × 200) and different DNA size markers (φX174-*Hae*III or φX 174-*Hinf* I). Use a 2% agarose horizontal gel for IL-1RN (VNTR) if desired.

7 Stain the gel with ethidium bromide, visualize on a UV-transilluminator, and take a permanent record of results with an instant camera or a CCD device and thermal printer. Care should be taken in handling and disposal of ethidium bromide, because of its potential carcinogenicity.

3.4 Specific notes for IL-1 and TNF DNA polymorphisms screening by PCR-RFLP

3.4.1 IL-1RN (VNTR)

The existence of a variable number of tandem repeats in intron 2 of the *IL1RN* gene was first reported during the cloning of the gene (42). This VNTR was characterized by Tarlow *et al.* (43) as a variable number (2–6) of 86 bp repeats.

- Gene accession number: X64532.

- *Oligonucleotide primers*:
  5′-CTC.AGC.AAC.ACT.CCT.AT-3′ (+2879/+2895)
  5′-TCC.TGG.TCT.GCA.GGT.AA-3′ (+3274/+3290)

- *Specific conditions*: Cycling is performed at [96 °C, 1 min] × 1; [94 °C, 1 min; 60 °C, 1 min; 70 °C, 2 min] × 35; [70 °C, 5 min] × 1; 4 °C. Electrophoresis in 2% agarose, 90 V, 30 min.

- *Interpretation*: The PCR product sizes are a direct indication of the number of repeats: the most frequent allele (allele 1) yields a 412 bp product. As the flanking regions extend for 66 bp, the remaining 344 bp imply four 86 bp repeats. Similarly, a 240 bp product indicates two repeats (allele 2), 326 is for three repeats (allele 3), 498 is five (allele 4), and 584 is six (allele 6). Frequencies in a North British Caucasian population for the four most frequent alleles are 0.734, 0.241, 0.021, and 0.004.

3.4.2 IL-1RN (+2018)

This single base variation (C/T at +2016) in exon 2 was described by Clay *et al.* (22). We designed these PCR primers (mismatched to the genomic sequence) to engineer two enzyme cutting sites on the two alleles. The two alleles are 100% in linkage disequilibrium with the two most frequent alleles of IL-1RN (VNTR).

- *Gene accession number*: X64532

- *Oligonucleotide primers*:
  5′-CTA.TCT.GAG.GAA.CAA.CCA.ACT.AGT.AGC-3′ (+1990/+2015)
  5′-TAG.GAC.ATT.GCA.CCT.AGG.GTT.TGT-3′   (+2133/+2156)

- *Specific conditions*: Cycling is performed at [96 °C, 1 min] × 1; [94 °C, 1 min; 57 °C, 1 min; 70 °C, 2 min] × 35; [70 °C, 5 min] × 1; 4 °C. Each PCR reaction is divided in two 25 μl aliquots: 5 Units of *Alu* I are added to one aliquot and 5 Units of *Msp* I to the other, in addition to 3 μl of the specific 10 × restriction buffer. Incubation is at 37 °C overnight. Electrophoresis is by PAGE 9%.

- *Interpretation*: The two enzymes, respectively, cut the two different alleles. *Alu* I will produce 126 + 28 bp fragments for allele 1, while it does not digest allele 2 (154 bp). *Msp* I will produce 125 + 29 bp with allele 2, while allele 1 is uncut (154 bp). Hence the two reactions (separated side by side in PAGE) will give inverted patterns of digestion for homozygote individuals, and identical patterns in heterozygotes. Allelic frequencies in a North British Caucasian population are 0.74 and 0.26. For 90% power at the 0.05 level of significance in a similar genetic pool, 251 cases should be studied to detect a 1.5-fold increase in frequency, or 420 cases for a 0.1 absolute increase in frequency.

3.4.3 IL-1A (−889)

The C/T single base variation in the IL-1A promoter was described by McDowell *et al.* (24). One of the PCR primers has a base change to create an *Nco*I site when amplifying allele 1 (cytosine at −889).

- *Gene accession number*: X03833
- *Oligonucleotide primers*:
  5'-AAG.CTT.GTT.CTA.CCA.CCT.GAA.CTA.GGC-3' (−967/−945)
  5'-TTA.CAT.ATG.AGC.CTT.CCA.TG-3'   (−888/−869)
- *Specific conditions*: MgCl$_2$ is used at a 1 mM final concentration, and PCR primers at 0.8 μM. Cycling is performed at [96 °C, 1 min] × 1; [94 °C, 1 min; 50 °C, 1 min; 72 °C, 2 min] × 45; [72 °C, 5 min] × 1; 4 °C. To each PCR reaction is added 6 Units of *Nco*I in addition to 3 μl of the specific 10 × restriction buffer. Incubation is at 37 °C overnight. Electrophoresis is by PAGE 6%.
- *Interpretation*: *Nco*I will produce 83 + 16 bp for allele 1, while it does not cut allele 2 (99 bp). Heterozygotes will have the three bands. Allelic frequencies in a North British Caucasian population are 0.71 and 0.29. For 90% power at the 0.05 level of significance in a similar genetic pool, 214 cases should be studied to detect a 1.5-fold increase in frequency, or 446 cases for a 0.1 absolute increase in frequency.

3.4.4 IL-1A (+4845)

This single base variation (C/T) in exon V was described by Gubler *et al.* in the cloning of human IL-1α (44) and reported again in a later paper (45). We have designed new PCR primers to create an *Fnu*4HI restriction site in allele 1 (A.G. Chaudhary, unpublished). We have found the two alleles to be in 100% linkage disequilibrium with alleles of IL-1A (−889) (46).

- *Gene accession number*: X03833
- *Oligonucleotide primers*:
  5'-ATG.GTT.TTA.GAA.ATC.ATC.AAG.CCT.AGG.GCA-3' (+4814/+4843)
  5'-AAT.GAA.AGG.AGG.GGA.GGA.TGA.CAG.AAA.TGT-3' (+5015/+5044)
- *Specific conditions*: MgCl$_2$ is used at a 1 mM final concentration, and PCR primers at 0.8 μM. DMSO is added at 5% and DNA template at 150 ng/50 μl PCR. Cycling is performed at [95 °C, 1 min] × 1; [94 °C, 1 min; 56 °C, 1 min; 72 °C, 2 min] × 35; [72 °C, 5 min] × 1; 4 °C. To each PCR reaction is added 2.5 Units of *Fnu*4H1 (NEB) in addition to 2 μl of the specific 10 × restriction buffer. Incubation is at 37 °C overnight. Electrophoresis is by PAGE 9%.
- *Interpretation*: The enzyme *Fnu*4H1 cuts a constant band of 76 bp (the absence of which indicates incomplete digestion) and two further bands of 29 and 124 bp with allele 1, or a single band of 153 bp for allele 2. Frequencies in a North British Caucasian population are 0.71 and 0.29.

3.4.5 IL-1B (−511)

This C/T single base variation in the IL-1 beta promoter was described in 1990 and published in 1992 (47).

- *Gene accession number*: X04500
- *Oligonucleotide primers*:
  5'-TGG.CAT.TGA.TCT.GGT.TCA.TC-3' (−702/−682)
  5'-GTT.TAG.GAA.TCT.TCC.CAC.TT-3' (−417/−397)

- *Specific conditions*: MgCl$_2$ is used at a 2.5 mM final concentration, and PCR primers at 1 μM. Cycling is performed at [95 °C, 1 min] $\times$ 1; [95 °C, 1 min; 53 °C, 1 min; 72 °C, 1 min] $\times$ 35; [72 °C, 5 min] $\times$ 1; 4 °C. Each PCR reaction is divided in two 25 μl aliquots: 3 Units of *Ava*I is added to one aliquot and 3.7 Units of *Bsu*36I to the other, in addition to 3 μl of the specific 10 $\times$ restriction buffer. Incubation is at 37 °C overnight. Electrophoresis is by PAGE 9%.

- *Interpretation*: The two enzymes cut, respectively, the two different alleles. *Ava*I will produce 190 + 114 bp for allele 1, while it does not cut allele 2 (304 bp). *Bsu*36I will produce 190 + 114 bp with allele 2, while allele 1 is uncut (304 bp). The restriction pattern obtained should be the inverse in the two aliquots (identifying homozygotes) or identical (heterozygotes). Frequencies in a North British Caucasian population are 0.61 and 0.39. For 90% power at the 0.05 level of significance in a similar genetic pool, 133 cases should be studied to detect a 1.5-fold increase in frequency, or 505 cases for a 0.1 absolute increase in frequency. We have found (Campbell and di Giovine, unpublished) the two alleles to be in 100% linkage disequilibrium with a recently described polymorphism at IL-1B (−32) (48).

3.4.6 IL-1B (+3954)

This polymorphism was described as a *Taq*I RFLP of IL-1 B (25). We have sequenced the most likely region implicated and found a C/T single base variation at +3954 in exon V which fully explains the RFLP. We have designed PCR primers to insert a control *Taq*I site, hence the product will contain one constant and one polymorphic restriction site for *Taq*I.

- *Gene accession number*: X04500

- *Oligonucleotide primers*:
  5′-CTC.AGG.TGT.CCT.CGA.AGA.AAT.CAA.A-3′ (+3844/+3868)
  5′-GCT.TTT.TTG.CTG.TGA.GTC.CCG-3′   (+4017/+4037)

- *Specific conditions*: MgCl$_2$ is used at a 2.5 mM final concentration, and DNA template at 150 ng/50 μl PCR. Cycling is performed at [95 °C, 2 min] $\times$ 1; [95 °C, 1 min; 67.5 °C, 1 min; 72 °C, 1 min] $\times$ 35; [72 °C, 5 min] $\times$ 1; 4 °C. To each PCR reaction is added 10 Units of *Taq*I (Promega) in addition to 3 μl of the specific 10 $\times$ restriction buffer. Incubation is at 65 °C overnight. Electrophoresis is by PAGE 9%.

- *Interpretation*: The enzyme cuts a constant band of 12 bp (the absence of which indicates incomplete digestion) and either two further bands of 85 and 97 bp (allele 1), or a single band of 182 bp (allele 2). Frequencies in a North British Caucasian population are 0.82 and 0.18. For 90% power at the 0.05 level of significance in a similar genetic pool, 408 cases should be studied to detect a 1.5-fold increase in frequency, or 333 cases for a 0.1 absolute increase in frequency. We have recently described a new marker in the IL-1B gene at position (+6912) (Campbell and di Giovine, unpublished) which is in 100% linkage disequilibrium with IL-1B (+3954).

3.4.7 TNFA (−308)

This single base variation (A/G) in the TNFA promoter was described by Wilson *et al.* in 1990 and published in 1992 (49). One of the PCR primers has a base change to create an *NcoI* site when amplifying allele 1. Frequencies in North British Caucasian population are 0.77 and 0.23.

- *Gene accession number*: X02910
- *Oligonucleotide primers*:
  5′-AGG.CAA.TAG.GTT.TTG.AGG.GCC.AT-3′ (−331/−309)
  5′-TCC.TCC.CTG.CTC.CGA.TTC.CG-3′ (−244/−226)
- *Specific conditions*: MgCl$_2$ is used at a final concentration of 1.5 mM, and PCR primers at 0.2 μM. Cycling is performed at [95 °C, 1 min] × 1; [94 °C, 1 min; 60 °C, 1 min; 72 °C, 1 min] × 35; [72 °C, 5 min] × 1; 4 °C. To each PCR reaction is added 6 Units of *NcoI* in addition to 3 μl of the specific 10 × restriction buffer. Incubation is at 37 °C overnight. Electrophoresis is by PAGE 6%.
- *Interpretation*: *NcoI* digestion will produce 87 + 20 bp for allele 1, while it does not cut allele 2 (107 bp). Heterozygotes will have the three bands. An alternative method of screening that was used in the original screening paper by Wilson *et al.* (36) uses single-stranded conformation polymorphism (SSCP) analysis. For 90% power at the 0.05 level of significance in a similar genetic pool, 297 cases should be studied to detect a 1.5-fold increase in frequency, or 391 cases for a 0.1 absolute increase in frequency.

3.4.8 TNFA (−238)

This single base variation in the TNFA promoter was described by D'Alfonso *et al.* in 1994 (50). One of the PCR primers has a base change to create an *AvaII* site when amplifying allele 1.

- *Gene accession number*: X02910
- *Oligonucleotide primers*:
  5′-GAA.GCC.CCT.CCC.AGT.TCT.AGT.TC-3′ (−425/−403)
  5′-CAC.TCC.CCA.TCC.TCC.CTG.GTC-3′ (−236/−217)
- *Specific conditions*: MgCl$_2$ is used at a final concentration of 2 mM, and PCR primers at 0.25 μM. Cycling is performed at [94 °C, 3 min] × 1; [94 °C, 1 min; 61 °C, 1 min; 72 °C, 1 min] × 35; [72 °C, 5 min] × 1; 4 °C. To each PCR reaction is added 5 Units of *AvaII* in addition to 3 μl of the specific 10 × restriction buffer. Incubation is at 37 °C overnight. Electrophoresis is by PAGE 12%.
- *Interpretation*: *AvaII* will produce a constant band of 77 bp, the absence of which indicates incomplete digestion. In addition to this, allele 1 will be digested as 63 + 49 + 21 bands, allele 2 as 70 + 63 bp. Heterozygotes will have a mixed pattern of restriction. Frequencies in a North British Caucasian population are 0.94 and 0.06. For 90% power at the 0.05 level of significance in a similar genetic pool, 1432 cases should be studied to detect a 1.5-fold increase in frequency, or 149 cases for a 0.1 absolute increase in frequency.

3.5 Fluorescence-based allelic discrimination

In population genetic studies the need for faster, reliable, and cost-effective methods is obvious: the number of tests in our laboratory is approximately 80 000 per annum and is still increasing. Although the PCR-RFLP methods described above remain as the reference (we routinely test about 10% of samples by both methods), we currently use high-throughput systems which allow one-step genotyping (no post-PCR processing) and handling of template and reactions in a 96-microwell format. Since 1994 we have been cooperating with PE-Applied Biosystems to apply the 5′ nuclease assay (TaqMan) (51) to test genotypes at the IL-1 and TNF loci.

The TaqMan assay utilizes the 5′–3′ nuclease activity of *Taq* DNA polymerase to cleave an allele-specific fluorogenic probe hybridized to the polymorphic site (52). This probe has been designed to have a primary fluorophor at the 5′ end (typically FAM or TET) and a quencher dye (usually TAMRA) at the 3′ end. Because of the physical proximity between the dyes, by FRET (Förster resonance energy transfer) the fluorescent signal upon activation will be at the wavelength of the quencher dye. However, cleavage of the labelled probe during the course of the assay results in an increase in fluorescence, which can be detected in the sealed tube or microtitre well. The fluorogenic probes are designed to anneal to the polymorphic region before the occurrence of complementary strand synthesis. When the *Taq* polymerase starts the elongation phase, its 5′ nuclease activity will dislodge or cleave any oligonucleotide (according to the energy of hybridization) encountered on the template strand. A probe that is annealed to the template but has a mismatch present (i.e. is not allele-specific) will be displaced, but cleavage will not occur—and consequently no increase or shift in fluorescence will occur. However, if it is cleaved (because of 100% hybridization), the 3′ dye will be diffused away, and the 5′ fluorophor will emit at its own wavelength upon excitation.

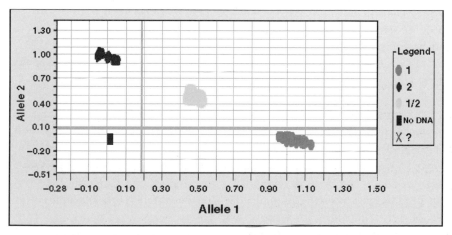

Figure 3 TaqMan genotyping. Shown are the clustering of samples according to their pattern of fluorescence at the end of a 5′ nuclease assay. Clustering at ± 2 SD from controls is detected and genotypes assigned accordingly.

35

By labelling the allele-specific probes with reporter dyes which fluoresce at different wavelengths, the reaction products will give different spectral profiles attributable to the genotype of the DNA samples from which they are generated. The ABI 'Sequence Detector Software' extracts the dye components from the overall spectra into separate values for the four dyes that are present in the assay. In addition to the reporter dyes and the quencher, a control dye (Rox) is also present in the reaction, where it plays no role apart from acting as an internal dye standard. The software performs a series of mathematical transformations on the data, with particular emphasis on a number of control samples of known genotype which are tested in every assay. The final transformation groups the samples together on the basis of their allele assignment, and genotypes are provided within a few minutes from the end of the PCR process.

A general protocol for TaqMan genotyping is provided below (*Protocol 6*). Sequences of probes, PCR primers, and PCR cycling profiles are detailed in Section 3.5.1. *Figure 3* illustrates typical TaqMan results.

Protocol 6

5′ Nuclease assay—TaqMan allelic discrimination

Equipment and reagents

- TaqMan Universal PCR Master Mix (P/N 4304437 from PE Biosystems), 2 × concentration. Optimized for TaqMan reactions, this contains AmpliTaq Gold DNA Polymerase, AmpErase UNG, dNTPs with dUTP and passive reference (ROX) in addition to optimized buffer components.
- DNA thermocycler with heated lid (PE-Biosystems, MJ Research)
- 96-well microwell plates, with optical caps (Advanced Biotechnologies)

- PE Biosystems Prism 7200, scanning fluorimeter
- PE Biosystems Sequence Detection System Software
- PE Biosystems Primer-Express Software
- Double-fluorescent oligonucleotide probes
- Control DNA templates: replicated control samples; homozygous allele 1 DNA; homozygous allele 2 DNA
- Template DNA
- FAM or TET labelled probes (PE Biosystems)

Method

1 Design probes and PCR primers. Note that this is a process requiring specific training —requiring attention to several parameters including primer compatibility, and adequate on/off energy of hybridization. Primers and probes are synthesized commercially.

2 Optimize the concentration of the probes by measuring and matching their spectral abilities at different wavelengths—the final concentration is usually optimal at 50–300 nM.

3 Optimize the PCR primer concentration (typically 100–300 nM) by performing serial PCRs and assessing the total measurable fluorescence.

4 Spot control templates (2 μl) into the wells of a microwell PCR plate. For each 96-well plate add: replicated control samples to 24 wells; no template (sterile water only) to 8 wells; homozygous allele 1 DNA to 8 wells; and homozygous allele 2 DNA to 8 wells.

5 Add the DNA samples (2 μl, 20 ng total) to the reaction plate.

6 Freshly prepare the reaction mixture as follows: TaqMan Universal Master Mix (1 × final concentration); PCR primers (50–900 nM each); FAM or TET labelled probes (50–300 nM final concentration); and 20 ng template DNA. (For a 96-well plate, it is typical to prepare sufficient mix for 100 reactions.) Dispense the reaction mix and mix to the template by pipetting using a multipipettor. Adjust the volume in each well to 23 μl with ddH$_2$O.

7 Seal the plate using either strips of optical caps or a clear heat-seal lid, centrifuge briefly to bring all the components to the bottom of the wells, and then place on a thermocycler fitted with a heated lid.

8 The cycling program is fixed for all TaqMan reactions—as probes and PCR primers have been designed to suit this profile in each reaction. Optimize the annealing temperatures to achieve maximum discrimination if necessary. Step 1: 50 °C, 2 min. Step 2: 95 °C 10 min. Step 3: 95 °C, 15 sec. Step 4: 57–64 °C (annealing), 1 min. Step 5: go to step 3, 40 times. Step 6: 15 °C, hold. End.

9 Remove the plate and read the fluorescence on the ABI Prism 7200. Note that the software analyses the increase in fluorescence of each sample in relation to the control samples and displays the normalized results in the form of a graph (see *Figure 3*).

10 Manually bin the genotypes into a database, following checks of spectral profiles.

3.5.1 TaqMan genotyping at the IL-1 and TNF locus

(i) IL-1B (−511)

Probe 1:   5′-C(•FAM)T CTG CCT CGG GAG CTC TCT(•TAMRA)-3′
Probe 2:   5′-C(•TET)T CTG CCT CAG GAG CTC TCT(•TAMRA)-3′
Forward: 5′-GTT TAG GAA TCT TCC CAC TT-3′
Reverse: 5′-TGG CAT TGA TCT GGT TCA TC-3′
Cycling: [95 °C for 2 min, 53 °C for 1 min, 74 °C for 1 min] × 2; [95 °C for 1 min, 53 °C for 1 min, 74 °C for 1 min] × 35; [94 °C for 1 min, 53 °C for 1 min, 74 °C for 5 min] × 3.

(ii) IL-1B (+3954)

Probe 1:   5′-A(•FAM)CC TAT CTT CTT TGA CAC ATG GGA TAA CGA T(•TAMRA)-3′
Probe 2:   5′-A(•TET)CC TAT CTT CTT CGA CAC ATG GGA TAA CGA T(•TAMRA)-3′
Forward: 5′-GCT CAG GTG TCC TCC AAG AAA TC-3′
Reverse: 5′-GTG ATC GTA CAG GTG CAT CGT-3′

Cycling: [95 °C for 2 min, 62 °C for 1 min, 72 °C for 1 min] × 2; [95 °C for 1 min, 62 °C for 1 min, 72 °C for 1 min] × 27; [94 °C for 1 min, 62 °C for 1 min, 72 °C for 5 min] × 3

(iii) IL-1RN (+2018)

Probe 1:   5'-C(•FAM)AA CCA ACT AGT TGC TGG ATA CTT GCA AG(•TAMRA)-3'
Probe 2:   5'-C(•TET)AA CCA ACT AGT TGC CGG ATA CTT GCA AG(•TAMRA)-3'
Forward: 5'-AAG TTC TGG GGG ACA CAG GAA G-3'
Reverse: 5'-ACG GGC AAA GTG ACG TGA TG-3'
Cycling: [96 °C for 1 min] × 1; [94 °C for 1 min, 63 °C for 1 min, 70 °C for 1 min] × 35; [63 °C for 5 min, 70 °C for 5 min] × 1.

(iv) IL-1A (+4845)

Probe 1:   5'-C(•FAM)AA GCC TAG GTC ATC ACC TTT TAG CTT CTT T(•TAMRA)-3'
Probe 2:   5'-C(•TET)AA GCC TAG GTC AGC ACC TTT TAG CTT CTT T(•TAMRA)-3'
Forward: 5'-ACC CCC TCC AGA ACT ATT TTC CCT-3'
Reverse: 5'-TGT AAT GCA GCA GCC GTG AGG TAC-3'
Cycling: [95 °C for 2 min] × 1; [94 °C for 1 min, 65 °C for 1 min, 72 °C for 1 min] × 40; [94 °C for 12 min, 65 °C for 2 min, 72 °C for 5 min] × 1

(v) TNFA (−308)

Probe 1:   5'-A(•TET) CC CCG TCC CCA TGC CC (•TAMRA)-3'
Probe 2:   5'-A(•FAM) AC CCC GTC CTC ATG CCC C (•TAMRA)-3'
Forward: 5'-GGC CAC TGA CTG ATT TGT GTG T-3'
Reverse: 5'-CAA AAG AAA TGG AGG CAA TAG GTT-3'
Cycling: [50 °C for 2 min] × 1; [95 °C for 10 min] × 1; [95 °C for 15 sec, 58 °C for 1 min] × 40; [15 °C, hold]

4 Notes

4.1 Quality controls

Incomplete digestion is the most common cause of mistyping in PCR-RFLP genotyping methods. Most of our protocols are based on a double-cut strategy, for which either a second restriction cutting site is used for digestion control on the diagnostic cleavage, or one enzyme cuts one allelic DNA form, and a different enzyme cuts the other allele. In this case, each reaction is the control for the other. PCR conditions are tested (and, if necessary, reoptimized) for each DNA preparation not performed in our laboratory. Template DNA quality is assessed by spectrophotometry and by gel electrophoresis.

The possibility of cross-contamination is very high in PCR-based techniques. Although the genotyping is physically separated from any laboratory where relevant cloned fragments are being handled, it is still possible to have PCR-product carryover from previous experiments (from lab. coats, hair, skin, etc.). A 'PCR-carryover prevention kit' is available from Roche Diagnostics. This is based on the UNG treatment of samples prior to PCR, which will cleave all dUTP-containing DNA. As all PCRs are performed using dUTP instead of dTTP, all previous PCR products, but not native templates, will be cleaved in this digestion step. This enzyme is inactivated by the first temperature ramping (94 °C) and therefore normal PCR can take place without UNG activity.

If laboratories do not use this system (which is expensive), there are stringent rules that can be used to reduce the risk of artefacts due to contamination (see below).

4.1.1 Prevention of contamination

- Laboratories are divided into **green** (pre-PCR) and red (post-PCR) areas.
- All laboratories have dedicated white coats, and workers are encouraged to change lab gloves as frequently as possible.
- **Green** laboratories have the most stringent requirements. Only goods coming from other green areas can enter, anything (equipment included) that leaves them cannot re-enter. These usually include a store-room, a 'sample reception' area, a 'clean DNA room' (where DNA extraction and PCR preparation are performed) and offices.
- **Red** laboratories have open access, but material and equipment can only move to other red areas or be disposed of in bags for autoclaving or incineration. Red areas are where PCR and electrophoresis take place.
- Results and images are stored in computer files and transferred to the offices by local network.

4.1.2 Detection of contamination

All PCRs carry 10% negative controls which are randomly placed within the experiment. These are routinely represented by water controls. In the case of dried blood spots (*Protocol 1*), negative controls are also represented by fragments (2–3 mm^2) of paper from the edge of the card. For human blood DNA preparations, murine T-cell lysates are extracted at the same time as each new batch of frozen blood, and the resulting DNA used as negative control.

4.2 Design of studies and analysis of results

Traditional parametric analyses (requiring the specification of a distribution and/or the mode of inheritance) have been used successfully to locate genes for monogenic diseases following simple Mendelian modes of inheritance. More commonly used in the genetic analysis of complex diseases are non-parametric methods, since these work independently of inheritance specifications and are generally more powerful than parametric methods when parameters are misspecified. The choice of analysis method depends on whether the investigator wishes to perform a whole genome screen or use a candidate gene approach, since certain methods are best suited to just one of these two approaches or to specific pedigree structures.

The following sections contain an outline of the most commonly used non-parametric methods of analysis and their suitability to the candidate gene approach.

4.2.1 Association studies

An allele at a certain locus is said to be associated with a disease if the frequency for that allele is significantly increased in the disease population over that of the normal healthy control population. True associations are due to linkage disequilibrium, where the disease-causing allele at the 'disease' locus remains on the same haplotype as those alleles which were present at closely flanking loci when the ancestral mutation occurred. Thus, the frequency of any allele on the 'disease haplotype' (including, of course, the disease allele itself) will be increased in the disease population. Recombination over extremely small distances is very low. However, as the time from the ancestral mutation increases, the distance over which linkage disequilibrium acts decreases, thus reducing the length of the 'disease haplotype'. It is therefore easier to detect association in young, isolated populations with a single founder-mutation effect where linkage extends over larger distances, than in large mixed populations.

At present, association studies are only suited to the candidate gene approach due to the small distances over which associations are detectable. In the future it is proposed that genome-wide association studies will be performed using several biallelic markers in every gene (53). Currently, however, this is not an available option.

Care must be taken when selecting the disease population in an association study, since spurious positive results may occur as an artefact of population admixture. It is usually advisable to investigate within a single ethnic group, since allele frequencies may vary between different groups. Similarly, if a control population is needed, it must be matched to the disease group for ethnicity, and ideally sex and age.

(i) Case-control studies

These types of studies can be performed for both qualitative and quantitative phenotypes. Obvious advantages of this approach include the ease of collection of large populations, the possibility of recruitment of patients with 'early disease' phenotypes, and the possibility of analysing late-onset diseases, where parental DNA may not be available.

- *Qualitative*: At the candidate gene locus, allele frequencies (or alternatively genotype frequencies) within the disease and control populations are calculated. The analysis is simple, comprising a $2 \times n$ contingency table (n denoting the number of categories, 2 for allele frequencies, or 3 for genotypes at a biallelic locus), upon which a Chi-square test may be used to determine whether the proportions differ significantly between the disease and control populations.

- *Quantitative*: When looking for a disease susceptibility allele, the individuals in both populations are first phenotyped quantitatively (usually the disease is classified as attaining a certain threshold value, therefore the unaffected controls are individuals falling below this). All individuals are then subdivided into the three (or more) genotypes. If an allele responsible for the inflated phenotype value of the diseased individuals exists, it would be expected that these individuals carry at least one copy of it. Thus the median of these genotype groups would be higher than those of the non-carrier groups. The non-parametric test involves testing for significant difference between the medians of the different genotype (or carriage) groups. This may be done via a Mann–Whitney test (for two groups) or a Kruskall–Wallis (for more than two groups), although several other tests also exist. In exactly the same way, this type of analysis may also be performed solely within the disease group to determine alleles for severity.

Extension to more than one locus

For qualitative traits the simple, one locus case-control analysis can be extended to one involving several loci (given a sufficient sample size). In a similar way, a larger contingency table can be calculated, with groups now corresponding to composite genotypes. As before, a Chi-squared statistic can be calculated. With these large contingency tables it is likely that the validity of the Chi-square test is violated ($< 80\%$ of expected values > 5, and expected values < 1). With smaller contingency tables, the usual remedy to violations of validity is to use Fishers Exact test, but in this larger case it is not viable. Instead a null distribution for the evaluated Chi-square statistic is simulated, and significance assessed from this. This test has been named the Monte Carlo Composite Genotype (MCCG) test.

(ii) Haplotype relative-risk (HRR) analysis

This analysis is only suitable for qualitative traits (quantitative traits may be used if dichotomized), and, as with all association tests, the candidate gene approach.

To perform an HRR analysis (54) parent/affected-offspring trios are needed. This type of analysis uses an artificial internal control, and therefore the problem of collecting an independent matched control population is removed. The parents and affected offspring are genotyped. It is then established which parental alleles were passed on to the affected offspring and which were not. From this the transmitted genotype and the non-transmitted genotype (internal control) are determined and recorded in the transmitted and non-transmitted groups, respectively. The two groups are then tested for significant differences in the proportions of their genotypes.

(iii) Transmission/disequilibrium test (TDT)

This analysis is suitable for qualitative traits investigated using a candidate gene approach. Parent/affected-offspring trios are needed, and if possible an unaffected sibling.

The TDT (55) is a test for both association and for linkage; more specifically, it tests for linkage in the presence of association. Thus, if association does not exist at the locus of interest, linkage will not be detected even if it exists. It is for this reason that the test has been included in this section. It may be used as an initial test, but is more commonly used when tentative evidence for association has already been identified. In this case a positive result will not only confirm the initial association, but also provide evidence for linkage.

Both parents and affected offspring are genotyped. Only parents heterozygous for the allele of interest may be used in the analysis. If the allele of interest is, or is linked to, the disease allele, the transmission rate for that allele from heterozygous parents to their affected offspring should be elevated. To test if the transmission rate of the allele of interest is significantly elevated, the number of times it is transmitted, b, and the number of times other alleles are transmitted, c, are counted. The squared difference of b and c divided by their sum provides a statistic that follows a Chi-square distribution with one degree of freedom, and can thus be assessed for significant deviation from the expected under no association or linkage. It is often advised to repeat this procedure using the unaffected offspring from the same parents to rule out the possibility of a spurious result due to biased meioses.

The TDT may also be used once linkage on a coarse scale has been shown to provide the fine-scale mapping that is necessary to pin-point the disease locus more accurately. Recently this type of methodology has been extended to quantitative traits (56) and late-onset traits where unaffected siblings can be used rather than parents (57).

4.2.2 Non-parametric linkage analysis

Non-parametric linkage analysis methods (such as Affected Sib-Pair analysis, the Haseman–Elston method, and the Variance Component Method) are based on the allele sharing status of affected relative pairs, usually sibs. These methods are suitable for whole genome screens (commonly done at 10 cM intervals) and also a candidate gene approach (although for fine localization alternative methods such as the TDT (see above) should be used).

Details of these methods are beyond the scope of this chapter, but can be found in the specialized literature (69–73).

4.2.3 Significance and power

Throughout this section, evidence strong enough to suggest association or linkage has been termed significant. The significance level of a test is left to the discretion of the investigator, but conventionally a 5% significance level is used. This means that it is accepted that there is enough evidence to suggest an association (or linkage) if the result would have occurred only 1 in 20 (0.05) times by chance in data where no association (linkage) existed—that is, there is only a 0.05 chance that the result is a false-positive. For each test a p-value may be calculated, which indicates the probability of the result occurring by chance.

In a single test, if this value is less than 0.05 then significant evidence may be claimed. This concept becomes more complicated when multiple, independent tests are performed. For example, if two tests were performed, and each was tested at the 5% level of significance, overall there is a 2 in 20 (0.1) chance of at least one result being a false-positive. Thus, for two independent tests, to maintain an overall significance level of 0.05 (0.05 chance of at least one test being a false-positive) either the individual significance level for each test must be lowered to 0.05/2 = 0.025, or the p-values doubled before assessing the result. This method of correction is called the Bonferroni correction. More generally, if n independent tests were carried out, each individual test should be tested at the 0.05/n level, or alternatively, every p-value multiplied by n before assessing the results. With non-independent tests, however, the Bonferroni correction may be too conservative.

Many investigators may find that they lose their potential significances through the dilution of p-values due to the correction criteria for multiple tests. Unfortunately these corrections are necessary for statistical correctness and cannot be discarded. However, if the results from the first set of observations are real, a second replication sample need only test those interesting results found from the first. This reduces the number of tests necessary on the second set of observations and thus reduces the dilution, increasing the chance of maintaining the statistical significance that may have been lost the first time. For complex diseases where there are so many questions to be answered it is perhaps unreasonable to expect that a single sample would be sufficient, and it would be wise instead to anticipate the necessity for a two-stage analysis and prepare accordingly. This is especially true for whole genome screens where the corrections necessary are massive. Lander and Kurglyak (58) list sensible guidelines for claiming significance in linkage analyses, specifically in the case of genome screens.

Along with significance, a second, and equally important, issue is that of power—the ability to pick up significant evidence where it actually exists. Given the phenotype, data structure, and number of observations, it is important to choose the method of analysis which is most likely to determine associations or linkages if they exist. In fact, it is advisable that in the planning stages of these studies the number of observations that are necessary to reach a predetermined power level are calculated. Unfortunately, this task is not as simple as it sounds, since power depends on several factors, of which some may be unknown, for example allele frequencies, marker informativeness, familial clustering of the disease, and recombination between marker and disease locus. Even if these factors are known, the power cannot be explicitly calculated for some methods, and instead empirical powers must be worked out via simulations.

There is no clear answer to which analyses should be done in different situations because of the many variables that are involved. However, it is strongly advisable to make the most informed choice possible, using previous work that has been done, to increase the chances of detection and location of the genes responsible, or involved in complex diseases.

References

1. di Giovine, F. S. and Duff, G. W. (1990). *Immunol. Today*, **11**, 13–20.
2. Beutler, B. and Cerami, A. (1989). *Annu. Rev. Immunol.*, **7**, 625–55.
3. Wordsworth, P. and Bell, J. (1991). *Ann. Rheum. Dis.*, **50**, 343–6.
4. Nicklin, M. J. H., Weith, A., and Duff, G. W. (1994). *Genomics*, **19**, 382–4.
5. Bird, T. A. and Saklatvala, J. (1986). *Nature*, **324**, 263–6.
6. Mitcham, J. L., Parnet, P., Bonnert, T. P., Garka, K. E., Gerhart, M. J., Slack, J. L., Gayle, M. A., Dower, S. K., and Sims, J. E. (1996). *J. Biol. Chem.*, **271**, 5777–83.
7. Seckinger, P., Williamson, K., Balavoine, J. F., Nach, B., Mazzei, G., Shaw, A., and Dayer, J. M. (1987). *J Immunol.*, **139**, 1541–5.
8. Eisenberg, S. P., Evans, R. J., Arend, W. P., Verderber, E., Brewer, M. T., Hannum, C. H., and Thompson, R. C. (1990). *Nature*, **343**, 341–6.
9. Carter, D. P., Deibel, M. R., Jr, and Dunn, C. J. (1990). *Nature*, **344**, 633–8.
10. Dinarello, C. A. (1991). *Blood*, **77**, 1627–52.
11. Saleeba, J. A., Ramus, S. J., and Cotton, R. G. (1992). *Hum. Mutat.* 1:63–69.
12. Ellis, E. A., Taylor, G. R., Banks, R., and Baumberg, S. (1994). *Nucleic Acids Res.*, **22**, 2710–11.
13. Fodde, R. and Losekoot, M. (1994). *Hum. Mut.*, **3**, 83–94.
14. Peeters, A. V. and Kotze, M. J. (1994). *PCR—Meth. Applic.*, **4**, 188–90.
15. Glavac, D. and Dean, M. (1993). *Hum. Mut.*, **2**, 404–14.
16. Blakemore, A. I. F., Tarlow, J. K., Cork, M. J., Gordon, C., Emery, P., and Duff, G. W. (1994). *Arthritis Rheum.*, **37**, 1380–5.
17. Mansfield, J. C., Holden, H., Tarlow, J. K., di Giovine, F. S., McDowell, T. L., Wilson, A. G., Holdsworth, C. D., and Duff, G. W. (1994). *Gastroenterology*, **106**, 637–42.
18. Tarlow, J. K., Clay, F. E., Cork, M. J., Blakemore, A. I. F., McDonagh, A. J. G., Messenger, A. G., and Duff, G. W. (1994). *J. Invest. Dermatol.*, **103**, 387–90.
19. Clay, F. E., Cork, M. J., Tarlow, J. K., Blakemore, A. I. F., Harrington, C. I., Lewis, F., and Duff, G. W. (1994). *Hum. Genet.*, **94**, 407–10.
20. Danis, V. A., Millington, M., Hyland, V. J., and Grennan, D. (1995). *Clin. Exp. Immunol.*, **99**, 303–10.
21. Mandrup-Poulsen, T., Pociot, F., Mølvig, J., Shapiro, L., Nilsson, P., Emdal, T., Røder, M., Kjems, L. L., Dinarello, C. A., and Nerup, J. (1994). *Diabetes*, **43**, 1242–7.
22. Clay, F. E., Tarlow, J. K., Cork, M. J., Cox, A., Nicklin, M. J. H., and Duff, G. W. (1996). *Hum. Genet.*, **97**, 723–6.
23. Tountas, N. A., Casini-Raggi, V., Yang, H., Kam, L., di Giovine, F. S., Rotter, J. I., and Cominelli, F. (1999). *Gastroenterology*, **117,** 806–13.
24. McDowell, T. L., Symons, J. A., Ploski, R., Forre, O., and Duff, G. W. (1995). *Arthritis Rheum.*, **38**, 221–8.
25. Pociot, F., Mølvig, J., Wogensen, L., Worsaae, H., and Nerup, J. (1992). *Eur. J. Clin. Invest.*, **22**, 396–402.
26. di Giovine, F. S., Cork, M. J., Crane, A., Mee, J. B., and Duff, G. W. (1995). *Cytokine*, **7**, 606–7.
27. Kornman, K. S., Crane, A., Wang, H. Y., di Giovine, F. S., Newman, M. G., Pirk, F. W., Wilson, T. G., Higinbottom, F. L., and Duff, G. W. (1997). *J. Clin. Periodontol.* **24**, 72–77.
28. Whyte, M., Hubbard, R., Meliconi, R., Timms, J., Duff, G. W., and di Giovine, F. S. (2000). *Am. J. Resp. Crit. Care* (in press).
29. Cox, A., Camp, N. J., Cannings, C., di Giovine, F. S., Dale, M., Worthington, J., John, S., Ollier, W. E. R., Silman, A. J., and Duff, G. W. (1999). *Hum. Mol. Genet.*, **8**, 1707–13.
30. Read, R. C., Camp, N. J., Borrow, R., Chaudhary, A., di Giovine, F. S., and Duff, G. W. (2000). *J. Infect. Dis.* (in press).

31. Cox, A., Camp, N. J., Nicklin, M. J. H., di Giovine, F. S., and Duff, G. W. (1998). *Am. J. Hum. Genet.*, **62**, 1180–8.

32. Svejgaard, A. and Ryder, L. P. (1989). *Genet. Epidemiol.*, **6**, 1–14.

33. Welch, T. R., Beischel, L. S., Balakrishnan, K., Quinlan, M., and West, C. D. (1988). *Dis. Markers*, **6**, 247–55.

34. Ahmed, A. R., Yunis, J. J., Marcus-Bagley, D., Yunis, E. J., Salazar, M., Katz, A. J., Awdeh, Z., and Alper, C. A. (1993). *J. Exp. Med.*, **178**, 2067–75.

35. Pociot, F., Briant, L., Jongeneel, C. V., Mølvig, J., Worsaae, H., Abbal, M., Thomsen, M., Nerup, J., and Cambon-Thomsen, A. (1993). *Eur. J. Immunol.*, **23**, 224–31.

36. Wilson, A. G., de Vries, N., Pociot, F., di Giovine, F. S., van der Putte, L. B. A., and Duff, G. W. (1993). *J. Exp. Med.*, **177**, 557–60.

37. McGuire, W., Hill, V. S., Allsopp, C. E. M., Greenwood, B. M., and Kwiatkowski, D. (1994). *Nature*, **371**, 508–11.

38. Copeman, J. B., Cucca, F., Hearne, C. M., Cornall, R. J., Reed, R. W., Rønningen, K. S., Undlien, D. E., Nisticò, L., Buzzetti, R., Tosi, R., Pociot, F., Nerup, J., Cornélis, F., Barnett, A. H., Bain, S. C., and Todd, J. A. (1995). *Nature Genet.*, **9**, 80–5.

39. Raskin, S., Phillips, J. A., Kaplan, G., Mcclure, M., and Vnencakjones, C. (1992). *PCR Methods Applic.*, **2**, 154–156.

40. Sambrook, J., Fritsch, E. F., and Maniatis, T. (1989). *Molecular cloning: a laboratory manual.* Cold Spring Harbour Laboratory Press, New York.

41. Ausubel, I. and Frederick, M. (1994). *Current protocols in molecular biology.* Wiley, New York.

42. Steinkasserer, A., Koelble, K., and Sim, R. B. (1991). *Nucleic Acids Res.*, **19**, 5095.

43. Tarlow, J. K., Blakemore, A. I. F., Lennard, A., Solari, R., Hughes, H. N., Steinkasserer, A., and Duff, G. W. (1993). *Hum. Genet.*, **91**, 403–4.

44. Gubler, U., Chua, A. O., and Lugg, D. K. (1989). In *Interleukin-1, inflammation and disease* (ed. R. Bomford and B. Henderson), pp. 31–45. Elsevier, Oxford.

45. Vandervelden, P. A. and Reitsma, P. H. (1993). *Hum. Mol. Genet.*, **2**, 1753.

46. Jouvenne, P., Chaudhary, A., Buchs, N., di Giovine, F. S., and Duff, G. W. (1999). *Eur. Cyt. Network*, **10**, 33–6.

47. di Giovine, F. S., Takhsh, E., Blakemore, A. I. F., and Duff, G. W. (1992). *Hum. Mol. Genet.*, **1**, 450.

48. Guasch, J. F., Bertina, R. M., and Reitsma, P. H. (1996). *Cytokine*, **8**, 598–602.

49. Wilson, A. G., di Giovine, F. S., Blakemore, A. I. H., and Duff, G. W. (1992). *Hum. Mol. Genet.*, **1**, 353.

50. D'Alfonso, S. and Richiardi, P. M. (1994). *Immunogenetics*, **39**, 150–4.

51. Livak, K. J., Flood, S. J. A., Marmaro, J., Giusti, W., and Deetz, K. (1995). *PCR–Meth. Applic.*, **4**, 357–62.

52. Holland, P. M., Abramson, R. D., Watson, R., and Gelfand, D. H. (1991). *Proc. Natl. Acad. Sci. USA*, **88**, 7276–80.

53. Risch, N. and Merikangas, K. (1996). *Science*, **273**, 1516–17.

54. Falk, C. T. and Rubinstein, P. (1987). *Ann. Hum. Genet.*, **51**, 227–33.

55. Spielman, R. S., McGinnis, R. E., and Ewens, W. J. (1993). *Am. J. Hum. Genet.*, **52**, 506–16.

56. Allison, D. (1997). *Am. J. Hum. Genet.*, **60**, 676–90.

57. Spielman, R. S. and Ewens, W. J. (1998). *Am. J. Hum. Genet.*, **62**, 450–8.

58. Lander, E. and Kurglyak, L. (1995). *Nature Genet.*, **11**, 241–7.

59. Udalova, I. A., Nedospasov, S. A., Webb, G. C., Chaplin, D. D., and Turetskaya, R. L. (1993). *Genomics*, **16**, 180–6.

60. Partanen, J. (1988). *Scand. J. Immunol.*, **28**, 313–16.

61. Jongeneel, C. V. and Cambon-Thomsen, A. (1991). *J. Exp. Med.*, **173**, 209–19.

62. Messer, G., Spengler, U., Jung, N. C., Honold, G., Blomer, K., Pape, G. R., Riethmuller, G., and Weiss, E. H. (1991). *J. Exp. Med.*, **173**, 209–19.

63. Ferencir, S., Lindemann, M., Horsthemke, B., and Grosse-Wilde, H. (1992). *Eur. J. Immunogenet.*, **19**, 425–30.

64. Bailly, S. (1993). *Eur. J. Immunol.*, **23**, 1240–5.

65. Todd, S. and Naylor, S. L. (1991). *Nucleic Acids Res.*, **19**, 3756.

66. Zuliani, G. and Hobbs, H. H. (1990). *Am. J. Hum. Genet.*, **46**, 963–9.

67. Sunden, S. L. F., Yandava, C. N., and Buetow, K. H. (1996). *Genomics*, **32**, 15–20.

68. Spurr, N. K., Hill, N., and Rocchi, M. (1996). *Cytogenet. Cell Genet.*, **73**, 256–68.

69. Sham P. C., Lin, M. W., Zhao, J. H., and Curtis, D. (2000). Power comparison of parametric and nonparametric linkage tests in small pedigrees. *Am. J. Hum. Genet.*, **66** (5): 1661–8.

70. Commenges, D., Beurton-Aimar, M. (1999). Multipoint linkage analysis using the weighted-pairwise correlation statistic. *Genet. Epidemiol.*, **17** Suppl 1: S515–9.

71. Alcais, A. and Abel, L. (1999). Maximum-Likelihood-Binomial method for genetic model-free linkage analysis of quantitative traits in sibships. *Genet. Epidemiol.*, **17** (2): 102–17.

72. McPeek, M S. (1999). Optimal allele-sharing statistics for genetic mapping using affected relatives. *Genet. Epidemiol.*, **16** (3): 225–49.

73. Gu, C., Province, M., Todorov, A., and Rao, D. C. (1998). Meta-analysis methodology for combining non-parametric sibpair linkage results: genetic homogeneity and identical markers. *Genet. Epidemiol.*, **15** (6): 609–26.

Chapter 3

Northern analysis, RT-PCR, RNase protection, cDNA array systems, and *in situ* hybridization to detect cytokine messenger RNA

Chris Scotton and Frances Burke

Biological Therapy Laboratory, Imperial Cancer Research Fund, PO Box 123, 44 Lincoln's Inn Fields, London WC2A 3PX, U.K.

1 Introduction

There are various methods used to detect cytokine mRNA in tissues and cell lines and the commonly used ones are described in this chapter. The varying degrees of sensitivity and suitability for use are discussed.

Northern analysis of RNA allows the quantitation of specific messenger RNA (mRNA) sequences and the determination of their size. The RNA sample is separated by electrophoresis through a denaturing agarose gel followed by capillary transfer to a membrane to which the RNA is subsequently covalently linked. The mRNA of interest is located by probing the membrane with a labelled complementary nucleotide sequence which, in the case of a radioisotopic label, is revealed by autoradiography. Reprobing for a suitable housekeeping gene allows quantitative comparison between samples on the same gel.

The ribonuclease protection assay (RPA) also allows the quantitation of specific RNA sequences. It has the advantage of being more sensitive than Northern analysis and many of the problems of non-specific binding are eliminated. We purchase RiboQuant® Multi-Probe RNase protection systems from Pharmingen to simultaneously assess the expression of several different messages. Template sets containing approximately 10 different probes have been successfully used in our laboratory. Using these template sets it is simple to analyse each band, and quantitation is made relatively straightforward due to the inclusion of housekeeping genes. The identities of the protected fragments are determined by plotting a standard curve of the length of each undigested probe versus the distance migrated through the gel. Using these kits, we have also been able to substitute ^{35}S for ^{32}P for the *in vitro* transcription, which is easier and safer to work with. A sequencing gel apparatus is required to separate the various RNase-protected fragments.

The reverse transcriptase-polymerase chain reaction (RT-PCR) is another powerful tool for analysing cytokine mRNA, with greater sensitivity than Northern analysis or RNase protection. It therefore allows the detection of cytokine mRNA when there are low copy-numbers per cell; in tissues where there are few expressing cells, or when only a small amount of sample is available. The method we use is non-quantitative and provides a positive or negative answer. Care must therefore be taken to include appropriate controls.

In situ hybridization (ISH) to cellular mRNA allows the analysis of gene expression in single cells. The availability of good immunohistochemistry techniques for cytokine detection is limited. This is in part due to the nature of these molecules; they are secreted extracellularly, rapidly diffuse from the site of secretion, and generally have a short biological half-life. The existence of membrane-bound forms and ubiquitously expressed receptors serve to confuse the identification of a secreting cell from a target cell from an 'innocent by-stander' by antibody detection methods. Therefore ISH is a useful technique for identifying cells expressing cytokine genes, though expression should obviously not be taken as an indicator of translation/secretion of the biologically active protein.

More recently, methods have become available to look at a large number of cytokines and their receptors in a given sample. In our laboratory, we have used cDNA arrays provided by Clontech with success. Although these kits are expensive, they are useful for providing answers to very specific questions. The kits we used provided us with two membranes each containing 268 cytokine/cytokine receptor cDNAs. We were able to look at the expression of these genes in untreated cells and cells that had been treated with IFN-γ. This is obviously useful in determining mechanisms of action of cytokines such as this. For this method the preparation of excellent quality RNA is essential.

2 Preparation and purification of RNA

2.1 General precautions for handling RNA

- Wear gloves and change them frequently throughout the experiment.
- Wash all glassware, spatulas, forceps, etc., rinse in alcohol, and then bake.
- Treat all buffers with 0.1% diethyl pyrocarbonate (DEPC) as follows: in a fume cupboard, add DEPC to 0.1% (v/v), shake well and leave at room temperature for at least 4 h, then autoclave (the exception here is Tris-based buffers where the DEPC reacts with amines in Tris). Prepare Tris buffers and alcohols in DEPC-treated distilled water.
- Wash gel tanks, combs, etc. in soap and water, followed by an ethanol or industrial alcohol rinse (do not wash the tray in ethanol as this will craze it). Treat with 3% (v/v) hydrogen peroxide in water for 10 min at room temperature, rinse well with DEPC-treated water and allow to dry.

2.2 Preparation of total RNA

There are now many commercially available reagents for the preparation of total RNA from cells and tissue. In our laboratory we routinely use Tri Reagent™ provided by Sigma, but there are other similar products available. The cells or tissue are lysed in a recommended volume of reagent which is essentially a phenol/guanidine mix. The protocol has the advantage of allowing the rapid preparation of a large number of total RNA samples. When using cell lines we have found it imperative that the manufacturer's guidelines are followed, and for adherent cells that the correct amount of reagent is used for the tissue-culture vessel surface area. This ensures that there is minimal protein and DNA contamination. In our laboratory we have found it practical to grow cell lines in triplicate in six-well, tissue culture dishes. At the time of lysis we then use 1 ml of Tri Reagent™ for three wells. This has enabled us to recover good quality total RNA and is cost-effective in terms of the amount of reagent we use. The method used is outlined in *Protocol 1*. This should be used in conjunction with the manufacturer's instructions which provide more detail.

Protocol 1

Isolation of total RNA

Equipment and reagents

- Tissue homogenizer (e.g. Ultra-turrax T25, Janke and Kunkel, Staufen, Germany) for tissues
- Microcentrifuge (4 °C)
- Spectrophotometer and quartz cuvettes (1 cm path length)
- 1.5 ml capped microcentrifuge tubes (Eppendorf)
- Dry ice/cardice
- Tri Reagent™ (Sigma)
- Chloroform (neat, without isoamyl alcohol)
- Isopropanol
- 75% ethanol
- DEPC-treated water

Method

NB: Tri Reagent™ is phenol-based and therefore must be used in the fume cupboard. Ensure that your hands/arms are completely covered.

Use 1 ml of Tri Reagent™ for 50–100 mg tissue. For adherent cell lines use 1 ml of Tri Reagent™ per 10 cm^2 of culture-plate surface area, or 1 ml for 3 wells of a six-well culture dish. For suspension cells, use 1 ml Tri Reagent™ for 5–10 $\times$ 10^6 cells. Scale-up accordingly.

1 Mince/grind frozen tissue on ice if necessary.

2 Homogenize tissue in 1 ml of Tri Reagent™ (clean the probe between each sample by washing first with ethanol then with DEPC-treated water).

3 For adherent cell lines, remove the culture medium, add Tri Reagent™ directly to wells, and pipette up and down to ensure lysis.

4 For suspension cells, spin down in a benchtop centrifuge, add Tri Reagent™ to cell pellet, and pipette up and down to ensure lysis. (The lysates can be left for approximately one month at $-70\,°C$ or used immediately.)

5 Transfer the lysates to microcentrifuge tubes and leave on the bench at room temperature for 5 min to allow dissociation of nucleoprotein complexes.

6 Add 0.2 ml of chloroform/1 ml of Tri Reagent™, vortex thoroughly, and leave on the bench at room temperature for 2–15 min.

7 Spin in a microcentrifuge at 10 000 g (13 000 r.p.m.) for 15 min at 4 °C.

8 Transfer the upper phase (ensure it is completely clear of the interface) to a fresh tube.

9 Add 0.5 ml of isopropanol/1 ml of Tri Reagent™. Invert to mix and leave on the bench for 5–10 min.

10 Spin in the microcentrifuge at 10 000 g (13 000 r.p.m.) for 10 min at 4 °C.

11 Decant the liquid. The RNA pellet should be clearly visible in the bottom of the tube.

12 Wash the pellet by adding 1 ml of 75% ethanol/1 ml of Tri Reagent™. Vortex well.

13 Spin in a microcentrifuge at 10 000 g (13 000 r.p.m.) for 10 min at 4 °C.

14 Decant the ethanol. Air-dry the RNA pellet for approximately 10 min. (Do not leave the pellet at room temperature for much longer than this as there is the danger of RNA degradation.)

15 Resuspend the pellet in an appropriate volume of DEPC-treated water. We usually resuspend in 30–100 μl of DEPC-treated water depending on the size of the pellet.

16 Determine the RNA quantity as follows:

For RNA an optical density of 1.0 is equivalent to 40 μg/ml or 0.04 μg/μl. Therefore in a 1 ml microcuvette with a path length of 1 cm, the following equation can be used to determine the RNA concentration:

$[RNA] = OD_{260} \times 0.04\ \mu g/\mu l \times$ dilution factor;

where $[RNA]$ = concentration of the RNA sample in μg/μl; OD_{260} = optical density at 260 nm; dilution factor = fold-dilution of RNA sample in the cuvette (see below).

For simplicity, the following method can be used:

(a) Take 4 μl of the RNA sample, add to 1 ml of DEPC-treated water (250-fold dilution, therefore the dilution factor is 250).

(b) Transfer the diluted sample to a quartz microcuvette, read the optical density (OD) at 260 nm.

(c) $[RNA] = OD_{260} \times 0.04\ \mu g/\mu l \times 250$.

Hence the concentration of the sample (μg/μl) will be $10 \times OD_{260}$.

For example, if the OD reading = 0.15 at 260 nm:

RNA concentration of the sample = 1.5 μg/μl.

17 To assess the purity of the sample, read the OD at 280 nm and calculate the ratio of $OD_{260}:OD_{280}$. If the ratio is 2.0 the sample is extremely pure, if the ratio is below

1.6 it is likely that the sample is contaminated by protein and it should therefore be re-extracted.

18 Resuspend the RNA at a useful working concentration. We routinely resuspend at 1 µg/µl in DEPC-treated water. Store at −70 °C for long-term storage or −20 °C for short term use. In this laboratory we have successfully kept RNA for 1 year at −20 °C without degradation.

Purification of mRNA from total RNA can be simply and rapidly performed using the numerous commercial kits now available. We have used the kit provided by Qiagen with success. Some kits extract mRNA directly from cell lines or tissues, whilst others recommend an initial purification of total RNA. Remember that mRNA is only 1–5% of the total RNA at the start of the experiment. In our hands we had to begin mRNA preparation with at least 500 µg of total RNA.

If enough starting material is available then (polyA)$^+$ mRNA rather than total RNA offers more sensitivity in Northern analysis. Typically, 2 µg of (polyA)$^+$ RNA is run per track. One disadvantage of purifying on the basis of the (polyA) tail is that it is often the first target for degradation and so may be lost in impure preparations of extractions from whole resected tissues.

For the purposes of RT-PCR, the total RNA must be of sufficient purity and integrity to allow efficient reverse transcription of the RNA to form cDNA. Of particular importance is the removal of contaminating genomic DNA to ensure that the specific PCR products are derived only from cDNA, and to avoid competition between the cDNA and genomic DNA during PCR. This can be achieved by DNase-treating the total RNA, as described in *Protocol 2*. We also recommend the use of citrate-buffered phenol (or water-saturated phenol) when precipitating RNA, since DNA is partitioned into the interphase or organic phase.

Protocol 2

DNase treatment of total RNA

Equipment and reagents

- Transcription optimized 5 × buffer (Promega)
- RNasin® ribonuclease inhibitor (Promega)
- 10 U/µl RNase-free DNase I (Pharmacia)
- DEPC-treated water
- Microcentrifuge
- 1.5 ml capped microcentrifuge tubes (Eppendorf)
- Absolute ethanol
- 75% ethanol
- 10 M ammonium acetate
- Citrate-buffered phenol (Sigma)
- Chloroform/isoamyl alcohol (24:1)
- Dry ice/cardice
- Dri-block
- 10 mg/ml carrier RNA (either tRNA or rRNA)

Protocol 2 continued

Method

1 In a microcentrifuge tube add, in this order:

   (a) 10 μl Transcription optimized 5 × buffer

   (b) 1 μl RNasin®

   (c) x μg total RNA (where x is between 1 and 20 μg of total RNA)

   (d) 2 μl DNase I

   (e) DEPC-treated water to a final volume of 50 μl

2 Incubate at 37 °C for 1 h.

3 Add 50 μl of DEPC-treated water.

4 (Optional) Add 0.5 μl of carrier RNA to aid the precipitation of small (< 10 μg) amounts of total RNA.

5 Add 100 μl of citrate-buffered phenol. Vortex thoroughly, then spin in the microcentrifuge at 10 000 g (13 000 r.p.m.) for 5 min at 4 °C.

6 Transfer the upper phase to a fresh tube. Add 100 μl of chloroform/isoamyl alcohol, vortex to mix, then spin in the microcentrifuge at 10 000 g (13 000 r.p.m.) for 5 min at 4 °C.

7 Transfer the upper phase to a fresh tube. Add 1/5 volume (i.e. 20 μl) of 10 M ammonium acetate and 2.5 volumes (i.e. 250 μl) of absolute ethanol. Put on dry ice for 30 min.

8 Spin in the microcentrifuge at 10 000 g (13 000 r.p.m.) for 15 min at 4 °C.

9 Remove the supernatant. The pellet should be visible at the bottom of the tube. Wash the pellet by adding 0.5 ml of 75% ethanol, then vortex.

10 Spin in the microcentrifuge at 10 000 g (13 000 r.p.m.) for 10 min at 4 °C. Remove the supernatant and air-dry the pellet.

11 Resuspend the pellet in an appropriate volume of DEPC-treated water. We usually resuspend the RNA at a concentration of 1 μg/μl.

3 Northern analysis

3.1 Electrophoresis of RNA for Northern blot

It is important to include positive controls of RNA extracted from cell lines that are known to express high levels of mRNA for the cytokine of interest.

For IL-1α, -β, and TNF-α, we use RNA extracted from HL60 cells stimulated with 50 ng/ml of phorbol myristate acetate (PMA); for IL-6 we use RNA from human foreskin fibroblast cells; for IFN-γ and IL-2 we use RNA from Jurkat cells stimulated with 1 μg/ml phytohaemagglutinin (PHA) and 50 ng/ml PMA for 8 h.

Membranes should be probed with appropriate housekeeping genes, such as β-actin or glyceraldehyde phosphate dehydrogenase (GAPDH), as controls for loading consistency and RNA integrity.

Protocol 3

Electrophoresis of RNA for Northern blot

Equipment and reagents

Caution: ethidium bromide is mutagenic, wear gloves to handle the gel, add hypochlorite to ethidium bromide solutions 30 min before discarding down the sink.

- Agarose (Ultrapure RNA grade)
- 10 × MOPS buffer (pH 5.5–7.0): 0.2 M MOPS (3-N-morpholinopropanesulfonic acid), 0.05 M sodium acetate, 0.01 M EDTA pH 8. Make up in deionized water and treat with DEPC as described earlier (Section 2.1).
- 37–40% formaldehyde solution (v/v)
- Northern loading buffer (aliquot and store at −20 °C): 0.72 ml deionized formamide, 0.16 ml 10 × MOPS buffer, 0.26 ml 37% formaldehyde, 0.18 ml DEPC-treated water, 0.1 ml 80% glycerol, 0.08 ml Bromophenol blue (saturated solution)

- 10 mg/ml ethidium bromide
- Deionized formamide
- 80% glycerol
- Horizontal gel electrophoresis apparatus
- Power supply
- Fume hood
- Dri-block
- Water bath or oven if used
- UV transilluminator and suitable equipment to capture the image
- Fluorescent ruler

Method

1 Prepare a 1% gel (if target mRNA is less than 2 kb use a higher percentage of agarose, e.g. 1.4%) as follows:

(a) 3 g agarose

(b) 30 ml 10 × MOPS

(c) 220 ml DEPC-treated water.

2 Microwave to dissolve the agarose then cool to 50 °C (this can either be done on the bench or by placing in a 50 °C oven/waterbath).

3 When cooled, add 50 ml of 37% formaldehyde in a fume hood, then mix gently.

4 Add 15 μl of 10 mg/ml ethidium bromide (this can be omitted if it is thought to interfere with transfer), mix and pour on to a levelled cleaned gel tray (see Section 2.1) with an appropriate comb fitted.

5 Allow the gel to set in a fume hood.

6 Remove the comb and any tape if used. Place the gel in an electrophoresis tank then cover with 1 × MOPS buffer diluted with DEPC-treated water.

7 Thaw RNA samples on ice (usually 15 μg), add 5 μl of loading buffer. Alternatively, use 2–5 μg of (polyA)$^+$ RNA. Heat the samples to 65 °C for 5 min, then put on ice prior to loading.

8 Run the gel at 90–100 V for approximately 4–5 h until the dye front has run 10–15 cm.

Protocol 3 continued

9   Photograph the gel on a UV transilluminator. Ribosomal RNA appears as two bright bands if ethidium bromide has been used in the gel. The upper band (28S) should be approximately twice as intense as the lower 18S band.

Protocol 4

Northern blotting (see *Figure 1*)

Equipment and reagents

- 20 × SSC: 3 M NaCl, 0.3 M Na$_3$ citrate·2H$_2$O, 800 ml distilled water Adjust to pH 7.0 with a few drops of 10 N NaOH, and make up to 1 litre
- DEPC-treated water (see Section 2.1)
- Whatman 3 MM filter paper
- Glass plates
- Reservoir tray
- Paper towels
- Nylon membrane at correct pore size, usually 0.45 μM, e.g. Hybond N (Amersham)
- UV cross-linker
- Saran wrap

Method

1   After photographing, wash gel in DEPC-treated water for 10 min followed by 30 min soaking in 20 × SSC. These steps will help to remove some of the formaldehyde and thus improve transfer. Place a glass plate across the reservoir tray filled with 20 × SSC.

2   Overlay the plate with a wick prepared from Whatman 3 MM paper, soaked in 20 × SSC, cut slightly wider than gel and long enough to drape into reservoir of 20 × SSC in the tray.

3   Smooth out any bubbles in the 3 MM wick using a pipette.

4   Place the gel on top of the wick and surround with Parafilm or suitable impermeable membrane to prevent 'short circuiting'.

5   Place the nylon membrane or equivalent (cut to size) on to the gel, again avoiding bubbles, and mark the membrane by cutting off one corner for subsequent orientation (check the manufacturer's instructions for pre-wetting requirements).

6   Cover the gel with two pieces of pre-wetted 3 MM paper cut to size.

7   Place a stack of paper towels on top (to form a layer approximately 10 cm or more thick).

8   Place another glass plate on top followed by a weight of approximately 500 g (e.g. a filled 400 ml bottle). Do not place too heavy an object on top as the gel may flatten prior to the transfer of all the RNA.

9   Leave overnight or for approximately 16 h to ensure complete transfer.

Protocol 4 continued

10 Recover the membrane, invert so it is RNA side up and dry on a piece of 3 MM filter paper for 5 min.

11 UV cross-link with 1200 J in a Stratalinker (always check the membrane manufacturer's instructions as to cross-linking procedures).

12 Either wrap the membrane in Saran wrap and store at 4 °C, or use immediately.

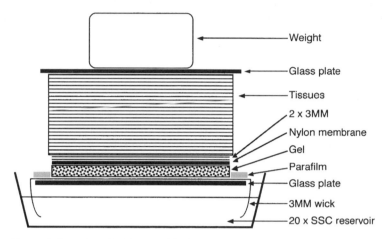

- Weight
- Glass plate
- Tissues
- 2 x 3MM
- Nylon membrane
- Gel
- Parafilm
- Glass plate
- 3MM wick
- 20 x SSC reservoir

Figure 1 Apparatus for Northern blotting.

3.2 Probe labelling

Protocol 5 is based on the random priming technique developed by Feinberg and Vogelstein (1, 2). In our laboratory we use a kit provided by Stratagene which allows the rapid labelling of the cDNA in approximately 15 minutes. Unincorporated radioactive precursors can be removed on one of many commercially available columns (e.g. Clontech make a range of columns with different exclusion sizes).

Protocol 5

Preparation of radiolabelled cDNA probes

Equipment and reagents

- Random priming kit (e.g. Stratagene Prime-It II kit for cDNA lengths of 200–1000 bp or Prime-It Tm for larger probe lengths)
- Dri-block at 37 °C and 100 °C
- Ice

- [^{32}P]dCTP 3000 Ci/mmol 10 MCi/ml (Amersham)
- Column (e.g. Clontech Chromaspin TE 100)
- Scintillation fluid
- Scintillation counter

Protocol 5 continued

Method

1 Label the probe by a random priming method according to the manufacturer's instructions.

2 Remove unincorporated nucleotide by passing through a suitable column (e.g. Clontech Chromaspin TE 100), again following the manufacturer's instructions.

3 Assess the labelling efficiency by counting 1 μl of the probe in an excess of scintillant (e.g. 8–10 ml). Expect more than 2×10^5 c.p.m./μl. Use approximately 1×10^6 c.p.m./ml of hybridization buffer (see *Protocol 6* for buffer).

4 Use the labelled probe immediately or store at 4 °C in a suitable shielded container.

5 Prior to use, heat the probe to 100 °C for 5 min, quench on ice for up to 30 min before adding to hybridization buffer. (Before boiling, ensure there is a hole in the top of the Eppendorf tube containing the labelled probe.)

Protocol 6

Pre-hybridization, hybridization, and post-hybridization of membranes

Equipment and reagents

- 1 M sodium phosphate buffer pH 7.2: make from stocks of 1 M Na_2HPO_4 and 1 M NaH_2PO_4 (approximately 2:1 mix)
- 0.1 M EDTA pH 8
- Bovine serum albumin (BSA)
- Sodium dodecylsulfate (SDS) (use a disposable face mask in addition to gloves when weighing out the powder)
- Post-hybridization wash buffer 1 ($\times$ 2) at room temperature (400 ml: 2 $\times$ SSC, 0.1% SDS)
- Post-hybridization wash buffer 2 ($\times$ 2) pre-heated to 68 °C (400 ml: 0.1 $\times$ SSC, 0.1% SDS)
- Post-hybridization wash buffer 3 at room temperature (400 ml: 2 $\times$ SSC)
- Ultrapure formamide
- Autoradiography film (e.g. Kodak XAR 5, Biomax MS), cassette, intensifying screens
- Hybridization oven/bottle system such as the Techne hybridizer
- X-ray film developer
- Waterbath (with shaking option)
- Saran wrap
- Vacuum grease
- Rotary mixer

Method

1 Preheat the hybridization oven to 42 °C prior to use.

2 Prepare the hybridization buffer as follows:

1 M sodium phosphate buffer pH 7.2	10.0 ml	0.2 M
0.1 M EDTA pH 8	0.5 ml	1 mM
BSA	0.5 g	1%

Protocol 6 continued

SDS	3.5 g	7%
Formamide	22.75 ml	45%
Distilled water	17.0 ml	34%
Total volume	50.0 ml	

Add in the following order to ensure the BSA and SDS dissolve easily: mix the phosphate buffer with the EDTA and water, then add the BSA and dissolve. Next add the formamide and finally add the SDS with gentle heating if necessary.

2   Place the UV cross-linked membrane (RNA side facing inwards) into a suitable hybridization tube which has been thoroughly cleaned prior to use. (Wash first with detergent followed by a distilled water rinse.) Ensure that the seals of the tubes cannot leak and apply a thin layer of vacuum grease to the rubber seal if necessary.

3   Add approximately 20 ml of the above hybridization mix. Pre-hybridization should take place for a minimum of 15 min at 42 °C. 30–60 min is optimal.

4   Replace with 20 ml of fresh hybridization mix with the probe added to a concentration of approximately 1×10^6 c.p.m./ml (heat the probe for 5 min at 100 °C to denature, quench on ice, then add to the hybridization mix prior to addition to the hybridization tube). Incubate overnight with rotation at 42 °C.

5   The following day pre-heat the waterbath and wash buffer 2 to 68 °C. Remove the filter from the roller. Wash the membranes in post-hybridization wash buffer 1, at room temperature for 5 min with shaking. Discard the buffer and repeat.

6   Discard the buffer and replace with post-hybridization wash buffer 2 pre-heated to 68 °C. Incubate in the shaking waterbath at 68 °C for 15 min. Discard the buffer and repeat.

7   Replace with post-hybridization wash buffer 3 at room temperature for 10 min.

8   Wrap the membrane in Saran wrap and expose overnight to X-ray film at −70 °C in a suitable cassette with intensifying screens. Adjust the exposure time as required. If you are planning to use densitometry ensure that the membrane is not overexposed.

9   Strip the membrane as follows: first rinse with boiling 0.1% SDS, discard and repeat, leave on a shaker until the solution cools down.

10 Wrap membrane in several layers of Saran wrap and store at 4 °C until required for reprobing. Ensure the membrane does not dry out after the first probing.

11 Use image analysis to look at relative levels of message when compared to housekeeping genes such as β-actin. A full description of this is outside the scope of this chapter refer to manual for Image analysis software e.g. 'NIH Image'.

The hybridization of the membrane described in *Protocol 6* is a modified version of that originally described by Church and Gilbert (3) for filter hybridization studies requiring high sensitivity and reprobing. There are numerous methods available now which allow hybridization to take place within an hour at 68 °C, in contrast to the overnight hybridization described in *Protocol 6*. In our experience we have found that we have been able to reprobe the membrane more often with the method outlined here. Using the quick hybridization methods we have only been able to reprobe approximately three times compared to six to seven times with the overnight incubation step.

4 RT-PCR for cytokine mRNA

PCR allows the rapid amplification of a DNA template, using specific oligonucleotide primers and a thermostable DNA polymerase. For the study of RNA, the sample of interest must therefore be converted into a DNA template. This is achieved through the use of reverse transcriptase (hence RT-PCR), which copies RNA to cDNA. Many kits are commercially available for cDNA synthesis, but in our laboratory we use the Ready-To-Go™ T-primed first-strand kit from Pharmacia Biotech. The method is outlined in *Protocol 7*, but refer to the manufacturer's instructions for more detail.

There are many different approaches to RT-PCR, and a variety of adaptations that allow semi-quantitation or comparison between different samples. In our laboratory we use RT-PCR only for qualitative purposes, deferring to Northern analysis or an RPA for a semi-quantitative determination of RNA levels. Therefore only basic RT-PCR will be described here.

Protocol 7

cDNA synthesis

Equipment and reagents

- DNase-treated total RNA
- 1.5 ml capped microcentrifuge tubes (Eppendorf)
- DEPC-treated water
- Ready-To-Go™ T-primed first-strand kit (Pharmacia)
- Dri-block
- Microcentrifuge

Method

1   Aliquot 5 μg of DNase-treated total RNA into a microcentrifuge tube, and adjust the volume to 33 μl with DEPC-treated water.

2   Incubate at 65 °C for 5 min.

3   Incubate the RNA sample at 37 °C for 5 min. At the same time, incubate a First-Strand Reaction Mix tube at 37 °C for 5 min, making sure that the vitrified pellet is at the bottom of the tube.

4   Transfer the RNA sample (33 μl) to the Reaction Strand Mix tube. **Do not** mix. Incubate at 37 °C for 5 min.

5   Mix the contents of the tube by pipetting the mixture up and down several times.

Take care to avoid clogging the pipette tip. Pulse-spin the tube in a microcentrifuge at 10 000 g (13 000 r.p.m.).

6   Incubate the tube at 37 °C for 1 h, to allow cDNA synthesis to proceed.

7   Adjust the volume of the sample to 50 μl by adding 17 μl of distilled water. Use 2 μl (equivalent to 200 ng of total RNA) for PCR (see *Protocol 8*).

8   Store at −20 °C until required.

Protocol 8

PCR reaction

Caution: ethidium bromide is mutagenic, wear gloves to handle the gel, add hypochlorite to ethidium bromide solutions 30 min before discarding down the sink.

Equipment and reagents

- cDNA (see *Protocol 7*)
- GeneAmp® 10 × PCR buffer (Perkin Elmer): 100 mM Tris–HCl pH 8.3; 500 mM KCl; 15 mM $MgCl_2$; 0.01% (w/v) gelatin; autoclaved
- 5 U/μl AmpliTaq® DNA polymerase (Perkin Elmer)
- Mixed GeneAmp® dNTP stock: 2.5 mM dATP, dCTP, dGTP, and dTTP (Perkin Elmer)
- Forward and reverse primers (100 μM stocks)
- DNA size marker (e.g. 123 bp marker, Gibco BRL, Paisley, UK)
- Light mineral oil (Sigma)
- Filter tips

- PCR tubes
- Thermal cycler (e.g. Perkin Elmer GeneAmp® PCR System 9700)
- Loading buffer: 40% (w/v) sucrose; 0.25% (w/v) Bromophenol blue; 0.25% xylene cyanol; make up in distilled water; autoclave
- UV transilluminator
- Submarine gel electrophoresis chamber
- Agarose
- 10 × TBE buffer stock: mix 108 g Tris base, 55 g boric acid, and 40 ml 0.5 M EDTA (pH 8.0). Add distilled water to a final volume of 1 litre.
- Ethidium bromide (10 mg/ml)

Method

1   Prepare a reaction mix containing the following:

GeneAmp® 10X PCR buffer	2.5 μl
dNTP mix	2.0 μl
Forward primer	1.0 μl
Reverse primer	1.0 μl
AmpliTaq®	0.2 μl (1 U)
Distilled water	16.3 μl

2   Aliquot 2 μl of the cDNA sample into a PCR tube. Include a negative control tube containing 2 μl of distilled water.[a]

Protocol 8 continued

3   Add 23 μl of the reaction mix to each tube, to give a final reaction volume of 25 μl.

4   Overlay with one drop of mineral oil[b] to prevent condensation and evaporation.

5   Place the tubes in the thermal cycler and proceed with the thermal cycling profile appropriate for the primers used.

6   Add 5 μl of loading buffer to 15 μl of the PCR product and resolve the products on a 1–1.2% agarose gel made with 1 × TBE and containing 0.5 μg/ml ethidium bromide. Bands can be visualized by UV transillumination and their sizes may be estimated with a co-migrated DNA size marker.

[a] A negative control is important to ensure that there has been no contamination of the reaction mix. Filter tips help to minimize the risk of contamination; we also recommend preparing PCR reactions in a laminar flow hood.

[b] Mineral oil can be omitted if the thermal cycler has a heated lid.

4.1 General remarks

The majority of primer pairs that specifically amplify cytokine or cytokine receptor cDNA are commercially available, or have been published. Primer design is crucial to the success of PCR and conditions for PCR amplification (e.g. annealing temperature, cycle number, and magnesium concentration) usually have to be optimized for each primer pair. Some parameters that may be altered are briefly described below. Further information on PCR optimization can be found in Dieffenbach *et al.* (4).

4.1.1 Primer design

PCR primers usually vary in length between 18 and 25 bases, and have a GC content of between 40 and 60%. For RT-PCR, primers are often designed to span introns thus preventing amplification of genomic DNA. In addition, primer sequences should be avoided that could result in the formation of primer dimers (particularly through complementary 3′ ends) or internal secondary structures.

We usually design primers that will anneal at a temperature of around 60 °C. However, the optimum annealing temperature can be determined empirically by performing the PCR reaction at several temperatures, until maximum specificity is achieved.

The primer concentration could also be optimized, by testing concentrations in the range of 0.1 to 1 μM.

4.1.2 Cycle parameters

A typical PCR cycle includes an initial melting step (94–96 °C), followed by a primer annealing step (40–65 °C), and then a primer extension step (72 °C). Normally, 30–35 cycles are sufficient for visualization of a product. In our laboratory, we routinely use the following thermal cycling profile: 94 °C (5 min); 35 cycles 94 °C (30 sec), 60 °C (30 sec), 72 °C (30 sec); 72 °C (7 min).

4.1.3 Magnesium concentration

The concentration of magnesium in the reaction buffer can affect the performance of the *Taq* polymerase. Insufficient free magnesium may inactivate *Taq* polymerase, while excess free magnesium can reduce the enzyme fidelity and result in non-specific amplification. The free magnesium concentration is also partially dependent on the dNTP concentration, since dNTPs can bind Mg^{2+}. If necessary, the magnesium concentration can be adjusted to between 0.5 and 3.5 mM.

4.1.4 Verification of PCR products

The identity of PCR products should be confirmed by Southern blotting or by direct sequencing of products extracted from the agarose gel. Once the specificity of the PCR primers has been demonstrated, this step can by omitted.

5 Ribonuclease protection assay

A number of ribonuclease protection assay (RPA) kits are now commercially available for the detection and quantitation of cytokine mRNA. In our laboratory, we use the RiboQuant® Multi-Probe RNase protection systems from Pharmingen: they have generated templates which are assembled into biologically relevant sets, allowing the user to simultaneously quantitate several different mRNAs in a single total RNA sample. A number of different template sets are available for the analysis of both human and murine cytokine and cytokine receptor mRNAs. They also provide template sets for investigating apoptosis, the cell cycle, etc.

The theory behind RPA is as follows: a radiolabelled antisense RNA probe is hybridized to target mRNA in the sample, forming double-stranded RNA, which is resistant to digestion by single-strand specific ribonucleases such as RNase A. After purification, the protected fragments are sized on a sequencing gel and can be quantitated by autoradiography and image-analysis or phosphoimaging. A radiolabelled antisense GAPDH or L32 probe added to each sample makes a good loading control, allowing for the semi-quantitation of each mRNA species in the original sample.

A representative autoradiograph for the hCR5 RiboQuant® Multi-Probe RNase protection system from Pharmingen, is shown in *Figure 2*.

5.1 General remarks

- Follow the precautions for handling RNA outlined in Section 2.1.

- Total RNA prepared according to *Protocol 1* is of suitable quality for use with the Pharmingen RPA kit. We normally use 5 μg of each total RNA sample.

- We have been able to substitute [α^{35}S]UTP (Amersham) for [α^{32}P]UTP in the *in vitro* transcription reaction. This is much easier and safer to work with, but requires longer exposure times during autoradiography.

- The use of Kodak Biomax MS film with a Kodak Transcreen LE intensifying screen (Sigma, UK) increases the sensitivity for low-energy emitters such as ^{35}S.

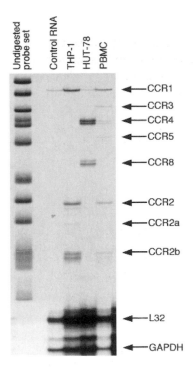

Figure 2 RNase protection assay using the RiboQuant® hCR5 template set from Pharmingen. The autoradiograph shows chemokine receptor expression by a monocytic cell line (THP-1), a T-cell line (HUT-78), and peripheral blood mononuclear cells (PBMC).

- Adhere strictly to the manufacturer's instructions for all other steps.

- In our laboratory, we have been using sequencing gel apparatus purchased from Bio-Rad for gel resolution of the protected fragments. This apparatus makes sequencing gel preparation much more straightforward than conventional methods.

6 *In situ* hybridization

The procedure can be divided into the following stages:

(a) rapid fixation/snap freezing of the sample (RNA degrades within minutes) and sectioning;

(b) pre-hybridization treatments to optimize probe access to target mRNA sequences, and reduce non-specific interactions;

(c) probe hybridization;

(d) post-hybridization washes to remove unbound probe reducing background signal;

(e) detection of bound probe.

For each tissue/cell sample outlined, methods may have to be adjusted to optimize cytological detail, probe hybridization, and signal-to-noise ratios.

6.1 Sample handling

Tissue samples such as biopsy specimens should be snap-frozen in isopentane (previously cooled on liquid nitrogen) immediately upon resection. Though we have found the use of frozen material more successful, wax-embedded tissue can also be used provided it is rapidly fixed in formal saline (time of fixation is dependent on the tissue type and sample size).

Peripheral blood mononuclear cells or cell suspensions should be cytospun immediately, but the slides can then be stored at $-70\,°C$ following fixation in 4% paraformaldehyde at room temperature for 5 min and dehydration through an alcohol gradient.

6.2 Pre-hybridization

These steps are used to optimize probe access and reduce backgrounds:

(a) Proteinase K treatment is included in the protocol for tissues but not for more fragile samples such as cytospin preparations. Controlled protease digestion partially removes cellular proteins to improve probe access.

(b) The amount of proteinase K required has to be optimized for each tissue (the range may vary widely from 1 μg/ml to 50 μg/ml), overdigestion will result in the loss of morphology.

(c) Treatment with 0.2 M HCl also provides an increase in signal, though certain samples may lose morphological detail.

(d) Acetylation with 0.25% acetic anhydride in triethanolamine reduces non-specific binding to the section and the slide by neutralizing electrostatic interactions.

6.3 Probe choice, labelling, and hybridization

Sensitivity of the *in situ* technique is dictated to a large extent by the type of probe used. Antisense asymmetrical probes have been used by many investigators to achieve good sensitivity with low backgrounds.

Advantages of asymmetric RNA probes:

(a) They contain only the sequences complementary to the target mRNA, whereas denatured double-stranded probes may reassociate during hybridization.

(b) Although DNA probes are more easily labelled by nick-translation or random priming, these labelling methods do not offer the specific activity of RNA probes labelled by the *in vitro* transcription protocol outlined later (see *Protocol 11*).

(c) Single-stranded (unbound) probe can be removed by incorporating a ribonuclease A digestion step in the post-hybridization stages thus reducing the background

(d) The thermal stability of RNA/RNA hybrids is greater than that of RNA/DNA hybrids, allowing increased hybridization and washing temperatures again serving to reduce non-specific binding.

The main disadvantage of using an RNA probe is the requirement for a sequence of interest to be cloned into a suitable vector. Most of the large molecular biology companies now provide kits to enable easy cloning and *in vitro* transcription. The sensitivity of RNA probes to degradation requires that suitable precautions are taken.

The increasing availability of oligonucleotides may provide an alternative to RNA probes offering resistance to degradation, single-strandedness, and tailored probe-length. Use of overlapping oligo 'cocktails' may provide good overall specific activity. It is widely accepted that short probe fragments (100–200 nucleotides) yield higher signals. In the case of RNA probes this can be achieved using controlled alkaline hydrolysis as outlined in the protocol.

Digestion time in the alkaline digestion buffer is calculated using the following equation:

$$t = \frac{(L_0 - L_f)}{kL_0L_f};$$

where t = time in min; L_0 = initial probe length in kb; L_f = desired probe length in kb; k = 0.11 (from Cox *et al.*, ref. 5).

The label of choice in this system is isotopic. The choice of isotope is ^{35}S, offering a compromise between low scatter (and therefore reasonable resolution) and a relatively short exposure time. [33]P is also widely used and requires shorter exposure times, although some reports suggest increased scatter.

Non-isotopic systems of labelling are improving all the time. Digoxigenin-based labelling systems are thought to be the more sensitive non-isotopic option. Boehringer–Mannheim (now Roche Molecular Biochemicals) supply easy to use DIG-labelling kits for preparing riboprobes.

6.4 Controls

The complexity of the technique demands rigorous controls: positive controls for the integrity of the probe and the integrity of the mRNA in the tissue; negative controls to verify that silver grain deposition in the emulsion is due to probe localization rather than, for example, noise or emulsion contamination.

6.4.1 Positive controls

Suitable probe controls would include tissues or cells known to contain cells expressing target mRNA (see Northern analysis controls). We also have excellent tissue controls in the form of cell lines transfected with the cytokine gene of interest established as tumours in nude mice. mRNA integrity in tissue can be verified by probing for a ubiquitously expressed message such as β-actin or GAPDH. Probes for polyadenylated sequences are likely to pick up partially degraded mRNA and, because of the vast difference in levels of abundance relative to cytokine mRNA, these would not be a relevant control.

6.4.2 Negative controls

Certain *in vitro* transcription vectors have paired RNA polymerase binding sites positioned on either side of the cloning site. These allow the generation of sense RNA probes from the insert. These will have comparable activity to the anti-sense probe yet will not bind to target sequences, thereby giving an idea of background levels. If these vectors are not available any irrelevant probe of similar activity may be used.

An idea of non-specific binding to tissue can be gained from pre-treating a small proportion of sections with ribonuclease A, in the pre-hybridization stages. The enzyme will degrade mRNA in the tissue section, preventing specific binding of the antisense probe.

Protocol 9

Slide preparation for *in situ* hybridization (ISH)

Equipment and reagents

- 400 ml glass staining troughs
- Slide racks
- Slides
- Coverslips
- 10% Decon-90
- Distilled water
- DEPC-treated water
- 3-aminopropyltriethoxysilane (TESPA), (Sigma)
- Acetone
- Sigmacote (Sigma)

Method

1 Prepare 10% Decon-90 detergent in distilled water.
2 Soak the slides (in racks) in 10% Decon for 30 min.
3 Rinse in running tap-water for 2 h.
4 Rinse in running distilled water for 2 h.
5 Allow the slides to dry, then bake overnight at 250 °C.
6 To improve adherence of tissue sections treat the slides as follows:
   (a) Make a solution of 2% TESPA (Sigma) in acetone.
   (b) Incubate the slides at room temperature for 10 sec.
   (c) Wash twice in acetone, then in DEPC-treated water.
   (d) Bake dry.
7 Siliconize coverslips by dipping in Sigmacote, dry dust-free, then bake.

6.5 Sample preparation

6.5.1 Solid tissue

Solid tissue (e.g. biopsy material) should be snap-frozen immediately in isopentane (previously cooled almost to its freezing point on liquid nitrogen) then stored at −70 °C prior to sectioning.

Cut 5 μm cryostat sections on to TESPA-coated slides using a new blade for each specimen. If this is impractical the blade should be frequently swabbed with alcohol. Cut sections may be stored in slide boxes with silica gel at −70 °C. Wax-embedded sections can be cut in a similar manner and run as for cryostat sections following standard dewaxing procedures.

6.5.2 Cells in suspension

Prepare cells in suspension, e.g. non-adherent cell lines or peripheral blood using a cytospin. Apply 100 μl of 1×10^6 cells/ml at 500 r.p.m. for 3 min. The cytospins are then fixed in 4% paraformaldehyde (PFA) for 5 min and either used immediately or dehydrated through an alcohol gradient and frozen at −70 °C until use. The procedure is then the same as is used for cryostat sections but treatment with proteinase K is omitted.

Cells may also be grown on coverslips and run as for cryostat sections.

Protocol 10

Pre-hybridization

Equipment and reagents

NB: All equipment and reagents used should be RNase-free. Buffers should be DEPC-treated as described in Section 2.1.

- DEPC-treated PBS
- 4% paraformaldehyde in DEPC-treated PBS, prepared fresh (dissolve at 80 °C in fume hood)
- 5–20 μg/ml proteinase K made up in appropriate buffer: Trizma-base, 5 mM EDTA pH 7.5
- DEPC-treated water
- 0.2 M HCl
- 0.1 M triethanolamine with 0.25% acetic anhydride added prior to use
- Alcohols: 30%, 50%, 70%, 95%, 100%, in DEPC-treated water
- Rotary shaker

Method

All incubations are carried out at room temperature on a rotary shaker. Place slides in racks, incubate as follows, in suitable slide troughs (e.g. 400 ml dishes (see *Protocol 9*))

1	PBS	5 min
2	DEPC-treated water	5 min
3	0.2 M HCl	20 min
4	PBS	5 min
5	4% PFA	15 min
6	PBS	2×5 min
7	5 μg/ml proteinase K	7.5 min
8	PBS	5 min
9	4% PFA	5 min

Protocol 10 continued

10	DEPC-treated water dip	
11	0.1 M triethanolamine + 0.25% acetic anhydride	2×10 min
12	PBS	5 min
13	Dehydrate 30%, 50%, 70%, 95%, 100%	2 min in each
14	Air-dry in a dust-free environment	

Protocol 11
Probe labelling

Most of the following components and *in vitro* transcription kits are available from Promega or Stratagene

NB: All equipment and reagents used should be RNase-free. Buffers should be DEPC-treated as described in Section 2.1.

Equipment and reagents

- $5 \times$ transcription buffer (see *Protocol 2*)
- 100 mM dithiothreitol (DTT)
- RNasin® ribonuclease inhibitor (Promega)
- 10 mM of each ATP, CTP, GTP
- $[\alpha^{35}S]UTP$, ^{33}P, or digoxigenin[a]
- Relevant RNA polymerase (e.g. T7, T3, SP6)
- RQ1 DNase (Promega)
- Ultrapure phenol
- 10 M ammonium acetate
- Absolute alcohol
- 10 mg/ml carrier RNA (either tRNA or rRNA)
- Dry ice/cardice

- Alkaline digestion buffer: 40 mM $NaHCO_3$, 60 mM Na_2CO_3 pH 10.2, add DTT to 10 mM before use
- 1 M sodium acetate
- 5% acetic acid
- Scintillation fluid
- Scintillation counter
- 1.5 ml microcentrifuge tubes, sterile and RNase-free
- DEPC-treated water
- Vacuum desiccator
- Dri-block

Method

1 To a sterile ribonuclease-free 1.5 ml microcentrifuge tube add the following:

(a)	$5 \times$ transcription buffer	4.0 μl
(b)	100 mM DTT	2.0 μl
(c)	RNasin	0.8 μl
(d)	ATP, CTP, GTP(mixed in a ratio of 1:1:1:1 with distilled water	4.0 μl
(e)	Linearized transcription vector template	1.0 μl (= 1 μg)
(f)	$[^{35}S]UTP$ (> 1000 μCi/nmol) or $[^{33}P]UTP$	10.0 μl
(g)	Relevant polymerase	1.0 μl

2 Pipette up and down gently to mix. Incubate at 37 °C for 1 h.

Protocol 11 continued

3 Add 2 units of RQ1 DNase, pipette up and down gently to mix.

4 Incubate at 37 °C for 30 min.

5 Add 80 μl of DEPC-treated water, phenol-extract (refer to *Protocol 2*), add 2 μl of 10 mg/ml carrier RNA, then precipitate with 20 μl of 10 M ammonium acetate and 300 μl of absolute alcohol.

6 Put on dry ice for 10 min. Microcentrifuge for 10 min at 4 °C. Carefully remove the supernatant.

7 Dry the pellet in a vacuum desiccator (or invert the microcentrifuge tube and dry in an RNase-free environment at room temp.) and take up in 100 μl of alkaline digestion buffer.

8 Incubate in a Dri-block or waterbath at 60 °C for 75 min (dependent on probe-length; see Section 6.3).

9 Stop this reaction with 10 μl of 1 M sodium acetate, 10 μl 5% acetic acid, and 2 μl 10 mg/ml carrier RNA.

10 Phenol-extract twice then precipitate as above (step 5), dry and resuspend in 40 μl of 10 mM DTT.

11 Count 1 μl in approximately 10 ml of a suitable scintillant (expect approximately $2–4 \times 10^6$ d.p.m./μl).

[a] For DIG-labelling follow the manufacturer's protocol.

Protocol 12

Hybridization

Equipment and reagents
Storage temperature are shown in brackets after each reagent
NB: All equipment and reagents used should be RNase-free. Buffers should be DEPC-treated as described in Section 2.1.

Equipment and reagents

- 1 M DTT (−20 °C)
- Deionized formamide (add 50 ml of formamide to 5 g of a mixed bed resin, Bio-Rad AG 501-X8, stir gently at 4 °C, filter, aliquot, and store at −20 °C)
- 100 × Denhardt's (4 °C)
- 1 M Trizma-base pH 8 (room temperature)
- 5 M NaCl (room temperature)
- 0.5 M EDTA (room temperature)
- 10 mg/ml (polyA) (−20 °C)

- 10 mg/ml carrier RNA, rRNA, or tRNA (−20 °C)
- 20 mM cold S-UTP (−20 °C) (may be omitted)
- 50% dextran sulfate (4 °C)
- 1.5 ml microcentrifuge tubes
- Siliconized coverslips (see *Protocol 9*)
- Sealable slide box
- 5 × SSC pH 7.0

Protocol 12 continued

Method

1 Prepare the hybridization solution by adding the following (μl volumes) to a 1.5 ml microcentrifuge tube. Note that 10–15 μl is required for each section. Final concentrations are given in brackets.

(a) 1 M DTT	5	(10 mM)
(b) Deionized formamide	300	(60%)
(c) 100 × Denhardt's	5	(1 ×)
(d) 1 M Tris pH 8	5	(10 mM)
(e) 5 M NaCl	30	(0.3 mM)
(f) 0.5 M EDTA	5	(1 mM)
(g) 10 mg/ml Poly A	15	(300 μg/μl)
(h) 10 mg/ml carrier RNA	15	(300 μg/μl)
(i) 20 mM cold S-UTP	11	(500 μM)
(j) 50% dextran sulfate	100	(10%)
Total volume	491	

2 Add the probe to a concentration of 5×10^4 c.p.m./μl.

3 Heat in a Dri-block for 2 min at 80 °C.

4 Apply 10–15 μl/section, cover with a siliconized coverslip.

5 Incubate overnight in a 50 °C incubator in a sealed slide box humidified with a tissue soaked in 50% formamide, 5 × SSC.

Protocol 13

Post-hybridization washes

Equipment and reagents

- 2-mercaptoethanol (2-ME)
- 5 × SSC
- 4 × SSC
- 2 × SSC
- 0.1 × SSC
- Formamide
- Ribonuclease buffer: 0.5 M NaCl, 10 mM Trizma-base, 5 mM EDTA, pH 8
- 10 mg/ml ribonuclease A
- Alcohols: 30%, 50%, 70%, 95%, 100%, in distilled water
- 0.1% gelatin (made up in distilled water)
- Slide racks
- 400 ml slide dishes
- Waterbaths at 37 °C, 50 °C, and 65 °C
- Autoradiograph equipment and facilities (see *Protocol 14*)

Method

1 Transfer slides to racks and incubate in the following buffers in 400 ml slide dishes equilibrated in 37 °C, 50 °C, and 65 °C waterbaths as directed. If possible, gently agitate the slides during the incubations:

(a)	5 × SSC, 0.1% 2-ME	50 °C	3 × 20 min
(b)	50% formamide, 2 × SSC, 0.1% 2-ME	65 °C	30 min
(c)	Ribonuclease Buffer	37 °C	2 × 10 min
(d)	Ribonuclease A, 20 μg/ml in the above	37 °C	30 min
(e)	Wash in (c)	37 °C	15 min
(f)	As (b)		
(g)	2 × SSC	room temp.	15 min
(h)	0.1 × SSC	room temp.	15 min

2 Dehydrate in 30%, 50%, 70%, 95%, 100% alcohol 2 min in each

3 Air-dry in a dust-free environment.

4 Dip slides in filtered 0.1% gelatin (made up in distilled water).

5 Air-dry in a dust-free environment.

6 Autoradiograph (see *Protocol 14*).

Protocol 14

Autoradiography

Equipment and reagents

- 0.1% gelatin (made up in distilled water, and filtered)
- Ilford K5 emulsion or equivalent
- Silica gel
- Kodak D-19 developer or equivalent
- 1% acetic acid
- 30% sodium thiosulfate (made fresh)

- Foil
- Toluidine Blue (1%)
- 50 °C waterbath
- Slide mailing box
- Light-proof box

Method

1 Pre-warm 10 ml of filtered 0.1% gelatin at 50 °C in a 50 ml measuring cylinder.

2 In a dark-room (suitable safelight may be used), top up to 20 ml with Ilford K5 emulsion.

3 Melt the emulsion in a 50 °C waterbath, rock gently (avoid bubbles) to mix.

4 Pour into a slide mailing box.

5 Hold the slides with forceps and dip into the emulsion, wipe the backs of the slides, then dry for 2 h in a light-proof box in a dark room.

6 Transfer the slides to a slide box containing silica gel then wrap the box in foil, put at room temperature overnight, then put to 4 °C for 7–10 days.

Protocol 14 continued

7   In a dark room, at room temperature, incubate slides in the following:

   (a) D-19 developer or equivalent for 2.5 min.

   (b) 1% acetic acid for 0.5 min

   (c) 30% sodium thiosulfate (freshly made) for 5 min.

8   Transfer the slides into distilled water then remove from the dark-room and wash in running distilled water tap for 60 min.

9   Counterstain with Toluidine Blue or other suitable general stain.

6.6 Interpretation

Probe localization will be detected under brightfield microscopy as a perinuclear deposition of silver grains over the cells expressing the target message, often giving a corona-like appearance (see Figure 3). If available, darkfield microscopy will accentuate this deposition but will also increase the background considerably under high magnification.

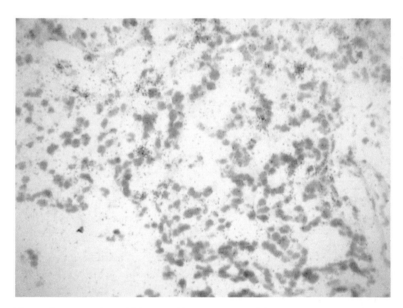

Figure 3 Localization of CC-chemokine receptor 1 (CCR1) by *in situ* hybridization to an human ovarian cancer biopsy section, using a ^{35}S-labelled antisense riboprobe.

Artefactual positives are common and may be due to several inapparent reasons, e.g. contamination of the emulsion with talc from gloves, precipitate in the gelatine, precipitation of components in the hybridization mix, or bubbles in the emulsion. Negative controls should be screened thoroughly and employed widely.

High backgrounds may be the result of many factors, for instance insufficient changes of washing buffer, insufficiently high temperatures, poor ribonuclease A digestion post-hybridization, or exposure to extraneous radiation during autoradiography.

References

1. Feinberg, A. P. and Vogelstein, B. (1983). *Anal. Biochem.,* **132**, 6.
2. Feinberg, A. P. and Vogelstein, B. (1984). *Anal. Biochem.,* **137**, 266.
3. Church, G. M. and Gilbert, W. (1984). *Proc. Natl. Acad. Sci. USA*, **81**, 1991.
4. Dieffenbach, C. W., Dveksler, G. S. (ed.) (1995). *PCR primer: a laboratory manual.* Cold Spring Harbor Laboratory Press, NY.
5. Cox, K. H., Deleon, D. V., Angerer, L. M., and Angerer, R. C. (1984). *Dev. Biol.,* **101**, 485.

Chapter 4

Purification, sequencing, and synthesis of cytokines and chemokines

Sofie Struyf, Anja Wuyts, and Jo Van Damme

Rega Institute for Medical Research, Laboratory of Molecular Immunology, University of Leuven, Minderbroedersstraat 10, B-3000 Leuven, Belgium

1 Introduction

This chapter describes the purification strategy routinely used in our laboratory to isolate and characterize novel cytokines from natural sources (cell cultures, body fluids) (1–3). The subsequent steps that constitute this strategy can easily be modified for the isolation of a particular cytokine by adapting the characteristics of the chromatography (type of columns and buffers) to the biochemical properties of the protein to be purified. The identification of cytokines through the isolation of natural proteins is a valuable approach in basic cytokine research. Based on relevant biological assays, new molecules can be isolated or new biological properties for a known cytokine might be revealed. For example, the isolation of the known chemokines RANTES, MIP-1α, and MIP-1β from conditioned medium, based on an assay that screened for factors protecting cells against HIV-1 infection, was the key discovery to the characterization of chemokine receptors as the co-receptors for HIV-1 and has opened up a new area of HIV-drug development (4). The availability of a pure protein allows its primary structure to be elucidated, which may eventually lead to the chemical synthesis or molecular cloning of the protein. In addition, physiologically relevant, post-translationally modified isoforms of the cytokine can be detected by this methodology (5). Finally, when an immunological assay is available to complement a less specific and more labour-intensive biotest during the purification procedure, naturally occurring inactive forms of a cytokine can also be identified. The availability of specific antibodies can also simplify the purification procedure of natural or recombinant proteins.

Figure 1 gives a general overview of the isolation strategy currently applied in our laboratory (6, 7). To obtain microgram amounts of natural cytokines, litres of conditioned medium (CM) from leucocytes or tumour cells are recommended as the starting material because cytokines and chemokines are generally produced

SOFIE STRUYF *ET AL.*

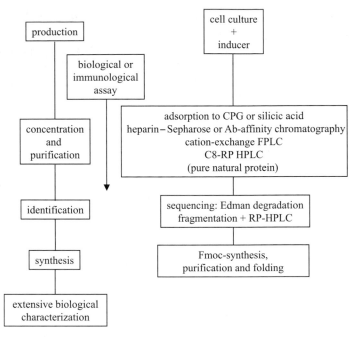

Figure 1 Schematic representation of the general strategy used to produce and purify cytokines for identification and characterization.

in minute amounts. Since the concentration of a recombinant cytokine produced by transfected cells is generally higher, smaller starting volumes might be sufficient. Substantial quantities of a particular cytokine will be produced only by certain cell types, providing that the cells are stimulated with an appropriate inducer. However, a mixture of natural cytokines is usually produced by cell cultures in response to a single or to multiple stimuli. If a specific immunological or biological assay is available, the best inducer or combination of inducers and their optimal concentrations should be determined first, as well as the timepoint at which the cytokine production reaches its maximum. Prior optimization of these parameters allows a reduction in the number of cell cultures to be handled.

The production of natural cytokines and chemokines can be stimulated by the addition of endogenous (cytokines) or exogenous (viral or bacterial products) inducers to the cell cultures. When *in vitro* cultured tumour cells are stimulated with a crude cytokine mixture, cytokine production normally reaches its maximum 48 h after induction. Production can also be maximized by restimulating cultured cells several times. Another important parameter that needs consideration is the amount of serum present in the induction medium. Since serum is a very rich and complex source of proteins, induction can best be performed in media with low serum concentrations. This facilitates the purification procedure and results in a homogeneous cytokine preparation. However, cytokine production levels under serum-free conditions are often rather low. After the initial concentration of CM, a cascade of chromatographic steps must yield a pure cytokine preparation that can eventually be identified by protein sequence analysis if sufficient amounts of pure protein are obtained. To reach that goal each individual purification step is optimized to maximize the recovery of biological activity. To trace the cytokine during the isolation procedure, a reliable biological or immunological assay is essential, allowing identification of those fractions which should be retained for further purification. If insufficient pure natural cytokine can be prepared for a detailed biochemical and biological characterization, molecular cloning or chemical synthesis is a valuable and fast alternative for generating larger amounts of the cytokine.

2 Purification of chemokines from natural sources

The chemokine family comprises about 50 structurally related, low molecular-weight (6 to 14 kDa) chemotactic cytokines, which selectively attract and activate various leucocyte subpopulations. Based on their production pattern, the members of this family are tentatively divided into inducible pro-inflammatory and constitutive chemokines. The first group of molecules attracts leucocytes mediating inflammatory reactions, whereas the second group regulates the normal homeostatic trafficking of immature blood cells and lymphocytes at various stages of maturation. In addition, chemokines and their receptors intervene in other physiological and pathological processes including HIV-infection, haemopoiesis, atherosclerosis, and angiogenesis.

In our laboratory natural chemokines are routinely isolated from *in vitro* cultured tumour cells or freshly isolated peripheral blood leucocytes using a four-step purification procedure (6, 7). The first isolation step concentrates the biological activity present in the large volume of CM. In the past, ultracentrifugation and precipitation of proteins by trichloroacetic acid, ethanol, or ammonium sulfate was often used as an initial concentration step. However, compared to batch adsorption to controlled pore glass (CPG) or silicic acid, poor purification and low recovery of biological activity were obtained by these methods, due to the instability of cytokines. Furthermore, the batch adsorption technique allows for a fast (hours) and sterile processing of unlimited volumes of CM, whereas a direct chromatographic approach takes much longer (days) thus resulting in a substantial loss of biological activity. Adsorption to CPG or silicic acid and elution at acidic or neutral pH results in a 10-fold reduction of the initial volume, with a good recovery of the biological activity (> 50%) if the technique (*Protocol 1*) is carefully executed.

Table 1 illustrates the purification of the CXC chemokine β-thromboglobulin by adsorption to silicic acid and assay of the chemotactic activity for neutrophilic granulocytes present in the CM of leucocytes. Since this procedure can yield a 50-fold reduction in total protein content, it also provides a first partial purification step (about 30-fold increase in specific activity). In order to check chemokine recovery in the eluate, chemotaxis assays can be performed on test samples, provided they are first dialysed to neutral pH and tested at several dilutions. Alternatively, detection of a single cytokine by a selective ELISA or by a more

Table 1 Purification and concentration of chemotactic activity and β-TG immunoreactivity from unfractionated peripheral blood leukocytes[a]

| Fraction | Volume (ml) | Protein (mg/ml) | Chemotactic activity[b] | | | β-Thromboglobulin[c] | |
			U/ml	Recovery (%)	Specific activity (U/mg)	μg/ml	Recovery (%)
Crude	6500	1.5	3.5	100	2.3	0.5	100
Unadsorbed	6500	0.8	<1.2	<34	<1.5	<0.1	<20
Elution							
—Ethylene glycol	270	0.3	53	62	175	6.6	55
—Acid pH	250	0.7	6.6	7.3	9.4	1.8	14

[a] Crude supernatant from leukocytes stimulated for 48 h with concanavalin A or LPS was stirred with 10 g/l silicic acid (2 h/4°C). Activities were recovered by elution with 50% ethylene glycol buffer, followed by elution with glycine-HCl pH 2.0 buffer (see *Protocol 1*). Values are averages of six different batches.

[b] Chemotactic activity for neutrophilic granulocytes determined by the agarose assay (8).

[c] β-Thromboglobulin concentration determined by radioimmunoassay.

Table 2 The multistep chromatography procedure to purify chemotactic cytokines

Purification step	Column	Pre-treatment Start buffer[a]	Elution buffer[a]	Gradient[a]	Flow rate	Fraction size
Heparin-affinity chromatography	heparin-Sepharose (Pharmacia)	dialysis 50 mM Tris–HCl 50 mM NaCl pH 7.4	50 mM Tris–HCl 2 M NaCl pH 7.4	0.05– 1M NaCl	20 ml/h	5 ml
Ab-affinity chromatography	CNBr-activated Sepharose 4B (Pharmacia)	dialysis PBS pH 7.4	0.5 M NaCl 0.1 M citrate–HCl pH 2.0	— (stepwise elution)	40 ml/h	5 ml
FPLC	Mono S cation-exchanger (Pharmacia)	dialysis 50 mM formate pH 4.0	50 mM formate 1 M NaCl pH 4.0	0–1 M NaCl	1 ml/min	1 ml
HPLC	C8 Aquapore RP-300 (PE Biosystems)	— 0.1% TFA pH 2.0	80% acetonitrile 0.1% TFA pH 2.0	0–80% acetonitrile	0.4 ml/min	0.4 ml
Gel filtration	Superdex 75 HR10/30 (Pharmacia)	concentrate to <150 μl 50 mM formate pH 4.0	50 mM formate pH 4.0	—	0.5 ml/min	0.25 ml

[a] In most purification steps, the pH and salt concentration of the input material must be readjusted by dialysis against the start buffer to maximize interaction with the matrix of the subsequent column. Depending on the type and concentration of the purified cytokine, 0.01% Tween-20 can be added to the dialysis/start and elution buffers to prevent loss of chemokine caused by adhesion to the dialysis membrane or collection tubes.

specific bioassay (e.g. hybridoma growth-factor activity for IL-6) can serve as a useful parameter for estimating the recovery. The chemokine recovery after adsorption to CPG is usually higher than that after adsorption to silicic acid, but CPG is much more expensive. Therefore, silicic acid is generally used in our laboratory for processing crude CM with a high protein content (e.g. body fluids or media supplemented with more than 2% serum). The relatively inexpensive silicic acid is discarded after use, whereas the expensive CPG is chemically cleaned (e.g. with nitric acid) and recovered. Following this initial concentration step, a series of different chromatographic purification steps (for an overview see *Table 2*) is applied to obtain a homogeneous chemokine preparation.

Protocol 1

Concentration and partial purification of crude conditioned media

Equipment and reagents

For Part A

- Spinner flask (Bellco)
- Controlled pore glass beads (CPG, particle size 120–200 mesh, pore size 35 nm; Serva)
- Phosphate-buffered saline (PBS)
- 10 mM glycine–HCl, pH 3.5
- Elution buffer: 300 mM glycine–HCl, pH 2.0
- Sintered glass filter (porosity 2 or 3, Schott)
- Dialysis buffer: 50 mM Tris–HCl, 50 mM NaCl, pH 7.4, containing 15% polyethylene glycol (average MW 20 000)
- Dialysis membranes with a molecular weight cut-off of 3.5 kDa (Spectra/Por, Spectrum Laboratories)
- 70% nitric acid in H_2O

For Part B

- Spinner flask or suspended stirring device
- Silicic acid (Matrex Silica, particle size 35–70 μm, pore size 10 nm; Millipore)
- 1 litre polypropylene centrifuge bottles (Nalgene)
- PBS
- PBS pH 7.4, containing 1 M NaCl
- Elution buffer: PBS pH 7.4, containing 1.4 M NaCl and 50% ethylene glycol
- Dialysis buffer and membranes (see Part A of this *Protocol*)
- Milli-Q H_2O (Millipore)

A. Batch adsorption to CPG and acid elution[a]

1 Transfer 1–3 litres of pre-cooled (4 °C) tissue culture CM to a spinner flask and adjust the pH to neutral to maximize CPG adsorption. Add 30 ml CPG per litre of CM. Place the suspension at 4 °C on a magnetic stirrer and allow adsorption of proteins by gentle stirring for 2 h. The optimal stirring speed is the lowest speed that keeps the CPG in suspension.

2 Turn off the stirrer and let the CPG settle down at 1 g (5 min). Carefully decant the supernatant and wash the CPG twice for 5 min by manual stirring, first with cold PBS (1/3 of the original volume of CM) and afterwards with 10 mM glycine–HCl, pH 3.5 (1/3 of the original volume).

Protocol 1 continued

3   Elute the adsorbed proteins by careful manual stirring for 5 min with cold elution buffer (1/30 of the original volume). Decant the eluate and repeat this manipulation once. Complete the elution process by magnetic stirring (twice, for 30 min each time at 4 °C) with fresh elution buffer (1/30 of the original volume). Pool the eluates after pouring over a sintered glass filter to remove CPG.

4   Before storage at −20 °C, further concentrate the CPG eluate by dialysis[b].

5   Recycle the CPG by removing strongly adsorbed proteins with 70% nitric acid in H_2O and by washing extensively with Milli-Q H_2O until neutral pH. Sterilize at 90 °C overnight and store CPG at 4 °C.

B. Batch adsorption to silicic acid and elution at neutral pH[c]

1   Add 10 g/l silicic acid to up to 10 l of pre-cooled CM at neutral pH and gently stir the suspension for 2 h at 4 °C.

2   Divide the suspension over centrifuge bottles, sediment the silicic acid by centrifugation (2000 g, 10 min, 4 °C) and carefully decant the supernatant. Wash the pellets first with cold PBS (1/3 of the original volume) and afterwards with 1 M NaCl in PBS (1/3 of the original volume).

3   Elute adsorbed proteins at neutral pH in cold elution buffer (1/20 of the original volume) by magnetic stirring for 30 min at 4 °C. Remove the eluate by centrifugation (2000 g, 20 min, 4 °C). Execute this elution four times in total to maximize the recovery and pool the centrifuged eluates for further processing.[d] Dialyse the eluate as described in Part A step 4, before storing at −20 °C.

[a] This method is suitable for 1–3 litres of CM with a low initial protein content (< 3 mg/ml); small volumes can be diluted to lower protein concentrations.

[b] Before application of the CPG or silicic acid eluate to the next purification step, the material needs to be dialysed to remove acid and salts that interfere with chemokine binding to the column matrix.

[c] This method is appropriate for large volumes of CM with a high protein concentration. Biological fluids (e.g. plasma or serum) can be pre-diluted in serum-free culture medium or PBS to lower the protein concentration and to maximize the adsorption.

[d] Normally, the bulk of cytokines adsorbed to the silicic acid is recovered by elution at neutral pH. However, an extra elution step at low pH (see Part A, steps 3 and 4) can be introduced to improve the yield for strongly adsorbed proteins. If only acid-stable proteins need to be recovered, elution in pH 2.0 buffer can be directly applied.

After the initial concentration step, the cytokine preparation is at an appropriate volume size to selectively isolate the cytokines by affinity chromatography. If specific antibodies are available for the cytokine to be purified, antibody-affinity chromatography is the first choice. Using a monoclonal antibody, pure cytokine preparations can theoretically be obtained after affinity chromatography. However, further processing of the eluate from the antibody column is often mandatory for final polishing. If novel or multiple natural chemokines need to be isolated, the common high affinity of these basic proteins for heparin is a useful characteristic, allowing separation of the chemokines from the majority of the proteins in the CPG/silicic acid eluate (see *Table 2*). After washing out weakly adsorbed proteins, chemokines are eluted from the heparin–Sepharose column in a broad range (0.5–2 M) by a NaCl gradient (0.05–2 M NaCl) due to their diverging affinity for heparin. Thus, a first fractionation of chemokines is achieved by this method, which allows a limited number of fractions to be selected for further purification. This selection is facilitated by biological (chemotaxis), immunological (ELISA), and biochemical (SDS-PAGE) testing of the individual column fractions. Based on their relative purity, a number of active peak fractions can be pooled for further purification. After heparin–Sepharose affinity chromatography, a more than 50-fold purification and 20-fold concentration can be achieved.

Subsequently, another characteristic of chemokines, i.e. their high isoelectric point, can be exploited for further purification by ion-exchange chromatography. Since the high salt concentration in the heparin–Sepharose fractions will disturb the interaction with the cation-exchange matrix, the selected fractions must be dialysed prior to loading (see *Table 2*). Moreover, dialysis adjusts the pH and the net charge of the chemokines. Elution of proteins from the cation-exchange column in a NaCl gradient (0–1 M) further removes contaminants and separates individual chemokines. At this stage, it is recommended that the proteins are eluted in a buffer containing detergent (0.01% Tween-20) to prevent pure (glyco)protein sticking to the collection tubes.

After this third purification step, SDS-PAGE analysis of the individual fractions often allows the proteins of interest to be visualized, providing they are present in sufficient amounts (1 μg/ml) and are devoid of numerous other low molecular weight proteins. However, after cation-exchange chromatography the numerous chemokines are usually nicely separated in the salt gradient. If the chemokine under investigation is eluting in a single sharp peak, another 5- to 10-fold purification and concentration is obtained by this method.

As a final purification step, high-pressure liquid chromatography (HPLC) is often chosen to separate proteins based on their hydrophobic interactions with the column matrix. Fractions eluting from the cation-exchange column can be directly injected on to a reversed-phase HPLC column (e.g. C8). After HPLC, homogeneous preparations of chemokines can be obtained, allowing for NH_2-terminal sequence analysis to elucidate their identity. An additional purification step is possible in case a high molecular-weight contaminant remains in the purified material, e.g. molecular sieving by gel filtration.

Each chromatography procedure consists of the same manipulations (conditioning, loading, washing, elution). The pH and salt concentration of the input material should be adjusted before loading it on to the column. For small volumes (1.5 ml), the buffer of a preparation can be exchanged by gel filtration or by desalting columns. However, in view of the larger elution volumes manipulated, we use dialysis against the start/equilibration buffer (see *Protocol 1* and *Table 2*). During dialysis, a precipitate might be formed, but it usually contains no bioactivity. Sterilization of the samples and buffers is recommended (0.22 μm filtration) before application, to increase the lifetime and performance of the columns. Before and after loading the preparation, the column should be equilibrated with a 'start' buffer (2–5 column volumes), at least until a stable baseline is reached. Details about the start and elution buffer composition, gradient, flow rate, and fraction size used for the different chromatographic steps are provided in *Table 2*. Between two purification runs, the column must be cleaned by a steep elution gradient (repeated two or three times) without application of a cytokine preparation.

All column fractions obtained by heparin–Sepharose chromatography, ion-exchange chromatography, and HPLC need to be checked for protein composition (SDS-PAGE) and biological potency (chemotaxis) to decide which fractions can be pooled for further processing. This important decision should be based on a balance between bioactivity and purity. In this respect it is acceptable to discard up to 20% of the total biological activity recovered, in favour of a higher purity. The protein composition of the fractions is routinely determined by SDS-PAGE on discontinuous Tris/tricine gels under reducing conditions as described in ref. 9. This method, which gives a superior resolution for proteins with a molecular weight in the 5–20 kDa range, is preferred because the molecular weight of chemokines varies between 6 and 14 kDa. The stacking, spacer, and separating gels contain 5% T (the total percentage concentration of both the acrylamide and bis-acrylamide monomers) and 5% C (the percentage concentration of the cross-linker relative to the total concentration T), 10% T and 3.3% C and 13% T and 5% C, respectively. In addition to the standard molecular weight markers for cytokines (ovalbumin, carbonic anhydrase, soybean trypsin inhibitor, lysozyme, purchased from Bio-Rad Laboratories) in each gel we include the low molecular weight (6500 Da) protein aprotinin (Pierce) as an additional marker for chemokines. Gels are stained by the very sensitive silver staining procedure that allows the detection of 10 ng of protein (10). At least 100 ng of chemokine is required to visualize the protein by classic Coomassie Blue-staining. This is important in view of the low abundancy of some cytokines and chemokines.

3 Complete identification of chemokines by amino-acid sequence analysis

If the purification strategy described above results in a pure cytokine preparation corresponding to the biological activity, NH_2-terminal amino-acid sequence analysis by automated Edman degradation can be performed to reveal the identity of the purified protein (see *Protocol 2*). The presence of conserved cysteine residues in the primary structure of chemokines cannot be directly detected, but is often obvious from the absence of any detectable signal after Edman degradation. Their presence can be confirmed by alkylation of the cysteines (see *Protocol 4*) (11, 12). The currently available automated protein sequencers have a high sensitivity, 10 pmol of protein being sufficient for the identification of 20 NH_2-terminal amino acids. Without fragmentation of the protein, a maximum of 50 residues can be determined with about 500 pmol of pure protein. Protein digestion is necessary to determine the full sequence of a peptide consisting of more than 50 amino acids. Peptide fragments can be generated either chemically or enzymatically (see *Protocol 3*). If more than one cleavage site for a selective proteinase is present in the peptide structure an extra HPLC purification step is mandatory to separate the resulting fragments prior to sequencing. The presence of a blocked NH_2-terminus, resistant to Edman degradation, is another reason to opt for protein fragmentation. For example, the monocyte chemotactic proteins MCP-1 to -4 all possess an NH_2-terminal pyroglutamate that is resistant to the first chemical reaction of the Edman degradation with phenylisothiocyanate. However, in the COOH-terminal part of these chemokines a single sequence Asp–Pro is present that can be chemically clipped by formic acid, allowing for immediate identification of the protein by sequencing its COOH-terminus after fragmentation. In case the NH_2-terminus is accessible, it can be blocked by *o*-phthalaldehyde (13), which also allows direct sequencing (without peptide separation by HPLC) of the COOH-terminal part if only a single formic-acid cleavage site is present. Selective cleavage of proteins by proteinases (e.g. Asp-N, Lys-C, Glu-C) normally yields multiple fragments even for small chemokines. NH_2-terminal sequencing of these individual fragments after fractionation by HPLC and the subsequent alignment of their sequences may result in complete identification of the primary protein structure.

Pure preparations are not always a prerequisite for sequence analysis. If contaminants are present in low molar concentrations compared to the cytokine of interest, its sequence can normally be distinguished from the weaker signal of the contaminating proteins. However, a better approach is to separate the contaminants in the preparation from the chemokine by SDS-PAGE. By subsequent electroblotting, the proteins are transferred to a polyvinylidene difluoride (PVDF) membrane and visualized by Coomassie Blue. The protein band corresponding to the chemokine can be excised and directly sequenced from the blot.

Protocol 2

Amino-acid sequence analysis[a]

Equipment and reagents

- Amino-acid sequencer (e.g. 477A/120A, PE Biosystems)
- Transfer buffer for semi-dry electroblotting: 50 mM Tris, 40 mM glycine, 0.04% SDS (w/v), 20% methanol (v/v) in H_2O, pH 9.2
- Milli-Q H_2O (Millipore)
- Semi-dry blot apparatus
- Polyvinylidene difluoride (PVDF) membrane (ProBlott membrane, PE Biosystems)
- ProSorb cartridge (PE Biosystems)

Method

1   Check the purity of the cytokine preparation by SDS-PAGE.[b] If contaminants are still present, than electroblot the sample after gel electrophoresis on to a PVDF

Protocol 2 continued

    membrane (1 mA/cm^2, 1.5 h) and visualize the proteins by Coomassie Blue staining. After destaining, rinse the membrane five times with Milli-Q H$_2$O and air-dry the membrane. Excise the protein band and load the strip on to the protein sequencer.

2   Bring the pure protein preparation in appropriate buffer[c] or desalt the sample prior to analysis using a commercially available desalting column (ProSorb Cartridge).

3   If the peptide is NH$_2$-terminally blocked or if an internal sequence is aimed at, clip the peptide using chemical cleavage or enzymatic digestion (see *Protocol 3*).

4   Pre-treat the protein with reducing/alkylating reagents if the peptide sequence is expected to contain cysteine residues (see *Protocol 4*).

5   Determine the NH$_2$-terminal amino-acid sequence on the automated (coupling, cleavage, and conversion) sequencer with on-line detection of phenylthiohydantoin (PTH) residues by HPLC.

6   Evaluate the sequencing data and determine the residue and yield for each step using an amino-acid standard.

[a] A whole volume of this Practical Approach series has been dedicated to protein sequencing (14).

[b] Discontinuous gel electrophoresis as described by Schägger and Von Jagow (9). The composition of the stacking, spacer, and separating gel is given in the text.

[c] The organic solvents used in the final HPLC purification step (acetonitrile in 0.1% TFA), like most organic solvents are compatible with amino-acid sequence analysis. Buffers containing salts or detergents, especially those containing free amino groups (e.g. Tris, glycine), should be avoided.

Protocol 3

Protein fragmentation for internal sequence analysis

Equipment and reagents

For Part A

- Formic acid
- C8 Aquapore RP-300 column (50 × 1 mm, PE Biosystems)
- 0.1% (v/v) TFA (HPLC grade) in Milli-Q H$_2$O (Millipore)
- 80% (v/v) acetonitrile (HPLC grade) in 0.1% TFA in Milli-Q H$_2$O

For Part B

- Sequencing-grade endoproteinase, e.g. Lys C (Roche Molecular Biochemicals))
- Milli-Q H$_2$O (Millipore)
- 2 × Digestion buffer,[a] e.g. for Lys-C: 50 mM Tris–HCl pH 8.5, 2 mM EDTA
- Dithiothreitol solution: 5 mg in 100 μl in Milli-Q H$_2$O
- C8 Aquapore RP-300 column (50 × 1 mm, PE Biosystems)
- 0.1% (v/v) TFA (HPLC grade) in Milli-Q H$_2$O
- 80% (v/v) acetonitrile (HPLC grade) in 0.1% TFA in Milli-Q H$_2$O

Protocol 3 continued

A. Chemical cleavage by formic acid

1 Concentrate the sample ($\geq$ 30 pmol) to 10–20 μl and add three volumes of 100% formic acid. Incubate the sample for 2 days at 37 °C. Choose one of the following steps (2–4) to continue the protocol depending on the number of expected cleavage sites in the protein.

2 If a single cleavage site Asp–Pro is present and the protein is NH_2-terminally blocked, directly load the digest on to the sequencer.

3 If one cleavage site is present in a protein that is not NH_2-terminally blocked and if only the sequence of the COOH-terminal fragment needs to be determined, block the original NH_2-terminus with *o*-phthalaldehyde before sequencing (13).

4 If multiple formic acid cleavage sites are present, separate the different fragments by RP-HPLC in a gradient (0–80%) of acetonitrile in 0.1% TFA in Milli-Q H_2O on a 50 $\times$ 1 mm C8 Aquapore RP-300 column (0.1 ml/min, 0.1 ml fractions).

B. Proteolytic digestion by selective enzymes

1 Concentrate the protein sample ($\geq$ 500 pmol) to about 15 μl. Add 60 μl of Milli-Q H_2O and 75 μl of 2 $\times$ digestion buffer and add the endoproteinase, e.g. Lys-C, at a concentration 1/20 to 1/100 to the protein substrate.

2 Incubate the mixture at 37 °C for 2–18 h.

3 Add 1 μl of the dithiothreitol solution and incubate at 37 °C for 1 h.[b]

4 Acidify the cleavage mixture and separate the peptide fragments by RP-HPLC as described in step 4 of Part A.

[a] The composition of the digestion buffer for each individual enzyme is provided by the manufacturer.

[b] Dithiothreitol reduces intramolecular disulfide bridges between the four conserved cysteine residues in chemokines, allowing separation of the protein fragments generated by the proteinase.

Protocol 4

Modification of cysteine residues for detection by sequence analysis

Equipment and reagents

For Part A

- Reduction/alkylation mixture: 2 μl of tri-*n*-butylphosphine and 5 μl of 4-vinylpyridine in 100 μl of acetonitrile (HPLC or Sequencer grade)

For Part B

- Reduction buffer: 200 mM Tris pH 8.4, 100 mM dithiothreitol, 1% SDS

- Milli-Q H_2O (Millipore)
- Alkylation reagent: 6 M acrylamide stock solution in Milli-Q H_2O
- Methanol (100% and 20% in Milli-Q H_2O)
- 0.1% TFA (HPLC or Sequencer grade) in Milli-Q H_2O
- ProSorb cartridge (PE Biosystems)

Protocol 4 continued

A. *In situ* alkylation

1 Load the protein sample ($\geq$ 30 pmol) on the sequencer cartridge, add 15 μl of the reduction/alkylation solution and reassemble the cartridge.

2 Equilibrate the reaction chamber with N-methylpiperidine, incubate for 70 min. Remove excess reagents by washing with n-butyl chloride and ethyl acetate by introducing an extra cycle in the sequencing program.

3 Continue with the normal reaction cycles.

B. Alkylation in solution

1 Lyophilize the protein ($\geq$ 100 pmol), solubilize in 10 μl of reduction buffer, and incubate for 30 min at 70 °C.

2 Dilute the protein sample with 40 μl of Milli-Q H_2O. Add 25 μl of the 6 M acrylamide stock solution and incubate for 45 min at 37 °C in the dark.

3 Add methanol to a final concentration of 10% and desalt the solution in a ProSorb cartridge: wet the PVDF membrane with absolute methanol; load the sample; wash the membrane with 20% methanol, once with 0.1% TFA and three times with Milli-Q H_2O.

4 Let the membrane air dry, punch it out of the cartridge, add 4 μl of methanol, allow to air dry, and transfer to the sequencer.

4 Chemical synthesis of chemokines

In general, cytokines and chemokines are produced in minute amounts by cultured cells. Thus, even if the *in vitro* production parameters (producer cell type, inducer(s), stimulation period) and the purification procedure are optimized, only small quantities (microgram level) of pure natural cytokine can be recovered from large volumes of CM. Therefore, the production and purification of natural protein is an expensive and time-consuming procedure when substantial amounts are requested for a detailed biological characterization. Although it remains difficult to perfectly reproduce the natural cytokine by molecular cloning or chemical synthesis (e.g. due to glycosylation), these two methods are widely accepted and routinely applied. In our laboratory, large-scale production by solid-phase peptide synthesis (15) was used for the production of several chemokines (6, 16). Chemical synthesis is a fast alternative for generating sufficient protein e.g. for the preparation of antisera. These methods are essential for developing specific immunotests and antibody-affinity columns, both of which are helpful in the isolation and characterization of natural chemokine variants and in studying gene regulation (5, 17). Since chemokines are relatively small proteins (70 to 80 amino acids) chemical synthesis can be accomplished within

SOFIE STRUYF *ET AL.*

one week and medium-sized quantities (milligram level) can be rapidly generated. The 9-fluorenylmethoxycarbonyl (Fmoc) chemistry was preferred to the *t*-butyloxycarbonyl (Boc) method, because this latter method requires extensive safety precautions.

During solid-phase peptide synthesis the growing peptide chain stays attached to a resin during the whole procedure, with synthesis performed from the COOH-terminus towards the NH$_2$-terminus. After linkage of the COOH-terminal amino acid to the reactive groups of the resin, each following residue is coupled one by one, so that the same sequence of chemical reactions is completed each time. This coupling reaction cycle comprises four steps: deprotection of the Fmoc-protected α-amino group of the last coupled residue; activation of the next amino acid; actual coupling; and capping of the free NH$_2$-termini of non-reacted peptides. After the last coupling reaction the Fmoc group is removed from the NH$_2$-terminal amino acid and the resin is washed extensively and dried on the synthesizer. Current Fmoc chemistry is a compromise between maximal coupling efficiency and yet relatively mild reaction conditions to avoid side-chain reactions. To prevent undesired degradative side reactions, the use of highly purified reagents and solvents is mandatory. Performing each linking reaction with high efficiency is very important, since even a small loss of 0.1% in each coupling results in a yield of only 47.5% when a 75-residue peptide is synthesized. In addition, the multiple incomplete peptides that accumulate and contaminate the intact peptide complicate the subsequent purification of the full-size chemokine to homogeneity. The coupling efficiency is improved by programming double coupling reactions when troublesome linkage is expected. Once the whole peptide chain is assembled, it is cleaved off the resin by acidolysis with TFA. In the same reaction, side-chain protection groups are removed. The reaction conditions and incubation time for the side-chain removal must be optimized according to the amino-acid composition of the synthesized peptide. Filtration separates the synthesized peptides from the resin. The protein is precipitated, washed, and lyophilized to be stored at 4 °C. In a subsequent step, the full-size peptide must be isolated and purified to homogeneity, e.g. by RP-HPLC. Pure intact protein in the column eluate can be identified by SDS-PAGE and sequence analysis. However, mass spectrometry is the most precise technique to detect side-chain reactions or incomplete deprotection. Once a pure preparation of the intact chemokine is obtained, disulfide bridges are formed in a mixture of oxidized and reduced glutathione. Even for different but related proteins, such as chemokines, the composition of the folding buffer and the incubation time must be tested on small samples of the protein to determine the best folding-reaction conditions. Dimerization by covalent bonding and disulfide bridge formation can be controlled by mass spectrometry. SDS-PAGE under reducing and non-reducing conditions can also reveal the presence of multimeric forms of synthetic protein. The final quality control consists of a comparison of the synthetic and natural protein in biological assays.

Protocol 5

Solid-phase peptide synthesis of chemokines using Fmoc chemistry

Equipment and reagents

- Automated peptide synthesizer (e.g. Model 433A, PE Biosystems)
- Amino acids with an Fmoc-protected α-amino group and protected side chains
    Asn, Cys, Gln: trityl
    Ser, Thr: t-butyl
    Asp, Glu: t-butyl ester
    Lys: t-butyloxycarbonyl
    Arg: 2,2,5,7,8-pentamethylchroman-6-sulfonyl (Pmc)
- Milli-Q H_2O (Millipore)
- 4-hydroxymethyl-phenoxymethyl-copolystyrene (HMP) resin
- Cleavage mixture: 0.75 g crystalline phenol, 0.25 ml 1,2-ethanedithiol, 0.5 ml thioanisole, 0.5 ml Milli-Q H_2O, 10 ml TFA
- Methyl t-butyl ether (MTBE)

- Medium-porosity glass filter (*Protocol 1*)
- Preparative RP-HPLC column: Resource RPC column (Pharmacia) or RP-HPLC 100 × 8 mm C18 Delta-Pak (particle size 15 μm; pore size 30 nm) column (Waters, Millipore) or 250 × 10 mm Aquapore octyl column (PE Biosystems)
- Analytic RP-HPLC column (220 × 2.1 mm C8 Aquapore RP-300, PE Biosystems)
- Acetonitrile
- TFA
- Folding buffer: 1 mM EDTA, 0.3 mM oxidized glutathione, 3 mM reduced glutathione and 1 M guanidium chloride in 150 mM Tris–HCl, pH 8.7

Method

1 Rank the individual amino acids and closely check the exact order. Program double couplings for residues where less efficient linking can be anticipated before starting the automated synthesis run.

2 After synthesis wash the peptide resin thoroughly[a] with N-methylpyrrolidone (NMP) and dichloromethane and then dry the resin on the synthesiser.

3 Prepare the cleavage mixture on ice and degas. Cleave off the peptide from the resin and simultaneously remove the side-chain protection groups by incubation in the cleavage mixture at room temperature for 90 min with continuous stirring.[b] Separate the synthesized chemokine from the resin by filtration through a medium-porosity glass filter. Precipitate and wash (4 times) the protein in cold MTBE. Dissolve the peptide in Milli-Q H_2O, lyophilize, and store at 4 °C until purification.

4 Dissolve the lyophilized protein in 0.1% TFA and separate the intact peptide from the other incorrect synthesis products by preparative RP-HPLC. Elute the peptides from the HPLC column in a linear acetonitrile gradient (0 to 80% acetonitrile in 0.1% TFA, 2 ml/min, 2 ml fractions).

5 Identify the intact peptide by SDS-PAGE, NH_2-terminal sequence analysis, and/or mass spectrometry.

6 Incubate the purified unfolded chemokine (100 μg/ml) in folding buffer containing

oxidized and reduced glutathione for 90 min at room temperature to form disulfide bridges.

7   Purify the folded peptide by analytical RP-HPLC (220 × 2.1 mm C8 Aquapore RP-300 column, 0 to 80% acetonitrile gradient in 0.1% TFA, 0.4 ml/min, 0.4 ml fractions).

8   Analyse the RP-HPLC fractions of folded chemokine by mass spectrometry and/or SDS-PAGE to control for purity and proper folding.

[a] To remove residual basic *N,N*-dimethylformamide, which inhibits acidolysis

[b] Thioanisole is added to accelerate the cleavage of Pmc groups from Arg residues. Phenol, water, and ethanedithiol are scavengers that minimize side-chain reactions.

References

1.   Van Damme, J., Van Beeumen, J., Opdenakker, G., and Billiau, A. (1988). *J. Exp. Med.*, **167**, 1364.

2.   Van Damme, J., Decock, B., Lenaerts, J.-P., Conings, R., Bertini, R., Mantovani, A., and Billiau, A. (1989). *Eur. J. Immunol.*, **19**, 2367.

3.   Proost, P., De Wolf-Peeters, C., Conings, R., Opdenakker, G., Billiau, A., and Van Damme, J. (1993). *J. Immunol.*, **150**, 1000.

4.   Cocchi, F., DeVico, A. L., Garzino-Demo, A., Arya, S. K., Gallo, R. C., and Lusso, P. (1995). *Science*, **270**, 1811.

5.   Proost, P., Struyf, S., Couvreur, M., Lenaerts, J.-P., Conings, R., Menten, P., Verhaert, P., Wuyts, A., and Van Damme, J. (1998). *J. Immunol.*, **160**, 4034.

6.   Wuyts, A., Proost, P., Froyen, G., Haelens, A., Billiau, A., Opdenakker, G., and Van Damme, J. (1997). In *Methods in enzymology*, Vol. 101 (ed. R. Horuk), p. 13. Academic Press, London.

7.   Proost, P., Wuyts, A., Conings, R., Lenaerts, J.-P., Put, W., and Van Damme, J. (1996). *Methods, a companion to methods in enzymology*, **10**, 82.

8.   Van Damme, J., Lenaerts, J.-P., and Struyf, S. (2000). In *Cytokine cell biology, a practical approach* (ed. F. Balkwill). IRL Press, Oxford (In press.)

9.   Schägger, H. and Von Jagow, G. (1987). *Anal. Biochem.*, **166**, 368.

10. Guevara, J. Jr, Johnston, D. A., Ramagali, L. S., Martin, B. A., Capetillo, S., and Rodriguez, L. V. (1982). *Electrophoresis*, **3**, 197.

11. Andrews, P. C. and Dixon, J. E. (1987). *Anal. Biochem.*, **161**, 524.

12. Brune, D. C. (1992). *Anal. Biochem.*, **207**, 285.

13. Brauer, A. W., Oman, C. L., and Margolies, M. N. (1984). *Anal.Biochem.*, **137**, 134.

14. Findlay, J. B. C. and Geisow, M. J. (ed.) (1989). *Protein sequencing, a practical approach*. IRL Press, Oxford.

15. Atherton, E. and Sheppard, R. C. (ed.) (1989). *Solid phase peptide synthesis, a practical approach*. IRL Press, Oxford.

16. Proost, P., Van Leuven, P., Wuyts, A., Ebberink, R., Opdenakker, G., and Van Damme, J. (1995). *Cytokine*, **7**, 97.

17. Van Damme, J., Proost, P., Put, W., Arens, S., Lenaerts, J.-P., Conings, R., Opdenakker, G., Heremans, H., and Billiau, A. (1994). *J. Immunol.*, **152**, 5495.

Chapter 5
Receptor binding studies

Lara Izotova and Sidney Pestka

Department of Molecular Genetics and Microbiology, UMDNJ–Robert Wood
Johnson Medical School, Piscataway, New Jersey 08854–5635, USA

1 Introduction

The interferons (IFNs) are a family of proteins which elicit a multitude of cellular
responses including antiviral, antiproliferative, and immunoregulatory activities.
There are four classes of interferons that have been isolated and characterized
(1–11). These are the leucocyte or interferon alpha (IFN-α), fibroblast or interferon
beta (IFN-β), immune or interferon gamma (IFN-γ), and interferon omega (IFN-ω).
Binding to a specific cell-surface receptor is a necessary, but not sufficient,
condition for cellular activation in the case of the IFN-γ system (12–16). IFN-α, IFN-
β, and IFN-ω share a common receptor complex that is distinct from the IFN-γ
receptor complex (5, 10, 11, 16, 17).

During the course of our studies of the IFN receptors, we developed a novel
procedure for radiolabelling ligands to very high activities (18–21). The method
is based on the fact that human (hu), murine (mu), rat, and even bovine IFN-γs
contain at least one recognition sequence for the cAMP-dependent protein kinase
from bovine heart. Thus, this has enabled us to phosphorylate several recombinant IFN-γs to very high activities by utilizing $[\gamma\text{-}^{32}P]$ATP and a commercially
available protein kinase. None of the phosphorylated proteins exhibited any
significant loss of biological activity when assayed in a standard cytopathic-effect
inhibition assay for interferon activity. Serine is the only amino acid phosphorylated. Murine and rat IFN-γ each have a single phosphorylation site (22), whereas
huIFN-γ has two phosphorylation sites (23). The phosphorylation sites reside at
the carboxy-terminal end of the proteins. Because of its nearly perfect identity
with huIFN-γ in the carboxy-terminal region, bovine IFN-γ likely also has two
sites of phosphorylation (22).

The labelling procedure was extended to proteins lacking an intrinsic phosphorylation site. By employing oligonucleotide-directed mutagenesis, phosphorylation sites were introduced in huIFN-αA (24, 25), huIFN-αB2-P (26), and huIFN-
αA/D(Bgl) (26). Several mutant proteins were isolated and each was capable of
being phosphorylated with no loss of biological activity. The $[^{32}P]$huIFN-αA-Pl
and huIFN-αB2-P proteins have been used extensively in our laboratory to study
the huIFN-α/β receptor. In addition, the coding sequences for monoclonal anti-

bodies were modified to contain phosphorylation sites for the cAMP-dependent protein kinase, casein kinase I, casein kinase II, and Src tyrosine kinase in experiments which demonstrated that the procedure could be generally used with other protein kinases and with other proteins (27–33). Recently, IL-10 and vIL-10 were modified to contain the cAMP-dependent protein kinase phosphorylation site and were effectively used for studies of receptor binding (L. Izotova, S. Kotenko, S. Saccani, and S. Pestka, unpublished observations).

Due to the very high labelling efficiency of the phosphorylation reaction (up to 10-fold higher than many iodination procedures) and high retention of biological activity, the ^{32}P-labelled IFNs have become invaluable reagents in our efforts to characterize and clone the IFN receptors. Binding studies allow the dissociation constant (K_d) of the ligand–receptor complex to be estimated, as well as the number of receptors per cell or per gram of protein in a preparation of membranes. From this type of information, a suitable source for purifying the receptor protein can be identified. We have also set up a translation system in which a binding assay is performed directly on *Xenopus laevis* oocytes 48–72 h after microinjection with mRNA (34). The oocyte binding assay has been used to evaluate various mRNAs for their relative content of IFN-γ receptor RNA, as well as to test the binding capability of putative receptor clones whose RNA was transcribed *in vitro* from vectors driven by the strong SP or T7 phage promoters (35). If the bound [^{32}P]IFN-γ is then covalently cross-linked to the receptor, the molecular size of the receptor can be estimated after SDS-polyacrylamide gel electrophoresis (PAGE) of the covalent complex and visualization by autoradiography.

This chapter will deal primarily with radiolabelling IFNs, the analysis of binding assays, and a procedure for covalently cross-linking labelled IFN to its receptor. Finally, some data on the cloning of the hu- and muIFN-γ receptors and a functional huIFN-α/β receptor will be presented. Although most of the data presented are from our studies of the interaction of IFN-γ and its receptor, the procedures are applicable to other ligand–receptor systems. Specifically, we have applied these procedures for labelling monoclonal antibodies to study their binding to tumour-associated antigens (27–33) and to the study of the IL-10 receptor.

2 Materials and reagents

- [γ-^{32}P]ATP, ≥ 5000 Ci/mmol (Amersham or New England Nuclear) is dried in a Savant Speed-Vac Concentrator prior to use.
- The catalytic subunit of bovine heart cAMP-dependent protein kinase (Sigma). This is prepared in 50 mg/ml (324 mM) dithiothreitol at 30 units/μl and stored in liquid nitrogen in small aliquots.
- Recombinant mu- and huIFN-γ purified as described previously (34, 36) and stored in liquid nitrogen in small aliquots. These were obtained from PBL Biomedical Laboratories, New Brunswick, New Jersey. The phosphorylatable huIFN-αA (24, 25), huIFN-αB2-P (26), and huIFN-αA/D(*Bgl*), huIFN-β, and hu-IL-10 can also be obtained from PBL Biomedical Laboratories.

- Disuccinimidyl suberate (DSS, from Pierce) prepared fresh at 18.3 mg/ml (50 mM) in dimethylsulfoxide.

- Large *X. laevis* females, purchased from NASCO (Fort Atkinson, WI).

- Nucleotide triphosphates (Pharmacia LKB) prepared at about 10 mM in distilled water, neutralized with KOH, sterilized by filtration, aliquoted, and stored at $-20\,°C$.

- Deoxynucleotide triphosphates (Pharmacia LKB) prepared at about 10 mM in distilled water, neutralized with Tris base, sterilized by filtration, aliquoted, and stored at $-20\,°C$.

- 7-Methyl-G(5')ppp(5')G (translation cap analogue, from Pharmacia LKB or New England Biolabs) prepared at 10 mM in sterile distilled water and stored at $-20\,°C$.

- GeneAmp DNA amplification kit (Perkin-Elmer Cetus).

- Red Module cDNA synthesis kit (Invitrogen).

- Sequenase 2.0 sequencing kit (US Biochemicals).

- Restriction endonucleases (Boehringer-Mannheim (now Roche Molecular Biochemicals), Pharmacia LKB, and New England Biolabs).

- T4 DNA ligase (Invitrogen).

- AMV reverse transcriptase, *E. coli* DNA ligase, T4 DNA polymerase, SP polymerase, and T7 polymerase (Promega).

- All other chemicals are of reagent grade.

3 Methods used in receptor binding studies

3.1 Cell culture

Grow tissue culture cells in appropriate media (Gibco or Sigma) containing 10% heat-inactivated fetal bovine serum and 50 µg/ml gentamicin sulfate at $37\,°C$ in 5% CO_2. For binding experiments, wash suspension cells once and then resuspend in fresh medium. For adherent cells, remove the medium, wash cells once with phosphate-buffered saline lacking Ca^{2+} and Mg^{2+} (PBS), release with trypsin–EDTA, wash, and finally resuspend in fresh medium. (Some receptors may be trypsin sensitive—see Section 6.)

The biological activity of the interferons can be assayed in a standard cytopathic-effect inhibition assay (37) with cells and challenge virus appropriate to the type and species of interferon being examined.

3.2 Interferon radiolabelling

The conditions of the phosphorylation reaction can vary somewhat depending on the desired specific activity of the material. The following protocol usually yields high levels of labelling.

Protocol 1

Interferon radiolabelling

Equipment and reagents

- [γ-^{32}P]ATP 10 mMCi/ml 5000 Ci/mmol (Amersham Pharmacia Biotech UK)
- Catalytic subunit of the cAMP-dependent protein kinase
- Reaction buffer: 20 mM Tris–HCl, pH 7.4, 100 mM NaCl, 12 mM MgCl$_2$, and 5–20 mM dithiothreitol. A stock solution is stored at $-20\,°C$ after sterile filtration

- Bovine serum albumin
- Dialysis equipment Spectra/Por 2.1 Biotech Regenerated Cellulose (RC). Dialysis membranes: 10 000 Dulton MWCO (Fisher Scientific Ltd)
- 10 mM sodium pyrophosphate, pH 6.7
- Liquid nitrogen

Method

1   Incubate about 0.5–1 μg of protein at $30\,°C$ for 60 min with 1 mCi of [γ-^{32}P]ATP and 15–60 units of the catalytic subunit of the cAMP-dependent protein kinase. The reaction volume of 60 μl should also contain 20 mM Tris–HCl, pH 7.4, 100 mM NaCl, 12 mM MgCl$_2$, and 5–20 mM dithiothreitol (depending on how much kinase is used).

2   Dilute the reaction mixture with 0.25–0.5 ml of a cold solution of bovine serum albumin (5 mg/ml) in 10 mM sodium pyrophosphate, pH 6.7. The sodium pyrophosphate is used to inhibit dephosphorylation.

3   Dialyse twice against 1 litre or three times against 250 ml of 10 mM sodium pyrophosphate, pH 6.7 at $4\,°C$ to remove the unincorporated [γ-^{32}P]ATP. Continue for at least 6 h before changing the solution. Store the labelled protein in small aliquots in liquid nitrogen.

Typically, the degree of phosphorylation has ranged from about 1000–4000 Ci/mmol (65–254 μCi/μg) for muIFN-γ, 630–6000 Ci/mmol (36–340 μCi/μg) for huIFN-γ, 890–2600 Ci/mmol (45–135 μCi/μg) for huIFN-αA-P1, 1111 Ci/mmol ($\sim$ 57 μCi/μg) for huIFN-αB2-P, and 1028 Ci/mmol ($\sim$ 53 μCi/μg) for huIFN-αA/ D(*Bgl*).

3.3 Binding of [^{32}P]muIFN-γ to tissue culture cells

A typical plot of bound c.p.m. versus free c.p.m. is shown in *Figure 1*. Transformation of the data according to the Scatchard method (38) should give rise to a straight line when *bound c.p.m./free c.p.m.* is plotted against *bound c.p.m.* (*Figure 2*). By converting the *bound c.p.m.* on the abscissa into a molar quantity of bound interferon, the slope of the plot is then equal to $-K_d^{-1}$. By determining the intercept on the abscissa in c.p.m., converting it to the number of molecules of interferon from the specific radioactivity of the [^{32}P]muIFN-γ, and dividing by the number of cells in the 50 μl sample analysed, one can estimate the number of IFN receptors per cell.

Protocol 2

Binding of [^{32}P]muIFN-γ to tissue culture cells

Equipment and reagents

- [^{32}P]muIFN-γ (see Section 2) and unlabelled muIFN-γ
- Costar Spin-X filter unit (optional)
- 1.5 ml polypropylene tubes
- Long 0.4 ml polypropylene microcentrifuge tubes
- 5–10% (w/v) sucrose in PBS

- Eppendorf micro centrifuge with horizontal rotor (model 5415C-spins up to 16 000 × g-Brickman Inc.).
- Cutting pliers
- Liquid scintillation counter, vials, and scintillant

Method

1 Prior to binding to cells, centrifuge the [^{32}P]muIFN-γ at 15 000 g for 15 min or filter (e.g. Costar Spin-X filter unit) for 1–2 min at 4 °C. This procedure removes protein aggregates which result from freezing and thawing the [^{32}P]muIFN-γ solution.

2 Prepare cells as described in Section 3.1 and resuspend in medium at a concentration of about 5×10^6/ml.

3 Typically, place 240 µl of cells into the first tube in each of two sets of 1.5 ml polypropylene tubes. The remaining tubes should contain 120 µl of cells. The first set of tubes contains untreated cells, whereas the second set of tubes contains cells which were treated with unlabelled muIFN-γ at a concentration of 1 µg/ml for 5 min prior to making the aliquots. The control should contain unlabelled ligand at a concentration about 100-fold greater than the K_d. In practice, at least a 100-fold excess over the concentration of the radioactive ligand is usually sufficient.

4 Add about 1–2.5×10^6 c.p.m. of [^{32}P]muIFN-γ to the first tube of each set and make twofold serial dilutions. Gently resuspend the cells every 15 min.

5 Following incubation at 22–24 °C for 70–90 min, layer an aliquot of 50 µl (in duplicate) from each tube over a 0.35 ml cushion of 5–10% (w/v) sucrose in PBS in a long 0.4 ml polypropylene microcentrifuge tube.

6 Centrifuge the tubes for 1 min at about 10 000 g in a horizontal rotor.

7 Freeze the tubes in liquid nitrogen and cut off the tips containing the cell pellets with cutting pliers.

8 Determine the radioactivity of the tips (bound c.p.m.) and the rest of the tube (free c.p.m.) in a liquid scintillation counter. Define specific binding as the difference between binding in the absence (total binding) and binding in the presence (non-specific binding) of excess unlabelled IFN.

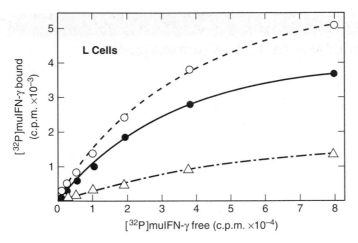

Figure 1 Concentration dependence of [^{32}P]muIFN-γ binding to mouse L cells. Binding to L cells (an adherent cell line) was carried out as described in *Protocol 2* and bound c.p.m. were plotted versus free c.p.m. for total binding ($\circ$), non-specific binding ($\triangledown$), and the calculated specific binding ($\bullet$, i.e. total minus non-specific binding). The data are taken from Langer *et al.* (21).

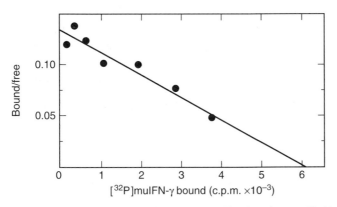

Figure 2 Scatchard analysis of binding data. The data for specific binding from *Figure 1* were plotted according to the Scatchard method (38). The slope of the resulting straight line is related to the K_d of the ligand–receptor interaction and the intercept on the abscissa is related to the number of receptors per cell. For L cells, several experiments gave a range of 3.0 to 6.9 $\times$ 10^{-10} M for the K_d and 6.0–16 $\times$ 10^3 receptors per cell. The data are taken from Langer *et al.* (21).

3.4 Binding of [^{32}P]muIFN-γ to *X. laevis* oocytes

The procedures for microinjecting RNA into frog oocytes as well as the binding assay have already been described in detail (34). Only the protocol for the binding assay will be repeated here for the sake of completeness.

Protocol 3

Binding of [³²P]muIFN-γ to *X. laevis* oocytes

Equipment and reagents

- Injected *X. laevis* oocytes (34, 35)
- 5 ml polypropylene tubes
- 0.5 × Leibovitz's L-15 medium (Gibco)
- Binding solution: 0.5 × L-15 medium (Gibco), 1 mg/ml bovine serum albumin, and 10^5 c.p.m. of [³²P]muIFN-γ (see Section 2)
- Unlabelled muIFN-γ

- Rotary shaker
- 1.5 ml polypropylene tubes
- Bray's solution (National Diagnostics)
- Ultrasonic processor (Model W-375, Heat Systems, Inc.)
- Liquid scintillation counter, vials, and scintillant

Method

1 Incubate injected oocytes at 18 °C for about 48 h. Place 10 oocytes (in duplicate or triplicate) in each 5 ml tube containing 200 μl of the binding solution. Set up another tube containing 10 oocytes, 200 μl of the same radioactive binding solution, and, in addition, 0.1 μg of unlabelled muIFN-γ as competitor. Clarify the binding solutions by centrifugation or filtration prior to use (see *Protocol 2*).

2 Incubate the oocytes at room temperature on a rotary shaker for 90 min.

3 Wash the oocytes three times with 3 ml of ice-cold 0.5 × L-15 medium as follows: add the medium, allow the oocytes to settle to the bottom of the tube, and then carefully remove the supernatant.

4 Transfer the oocytes to 1.5 ml polypropylene tubes and add 1 ml of Bray's solution.

5 Disrupt the oocytes in an ultrasonic processor and measure the associated radioactivity in a liquid scintillation counter (see an example of results in *Table 1*).

Table 1 Binding of [³²P]MuIFN-γ to cells and oocytes

Cell type	[³²P]MuIFN-γ bound (c.p.m.)		Ratio
	+	−	−/+
ABPL4 cells	1608	57 720	35.9
HL60 cells	1536	984	0.6
Injected oocytes	157	3000	19.0
Uninjected oocytes	285	312	1.1
Buffer-injected oocytes	178	211	1.2

The data in the table are expressed as c.p.m./1.5 × 10^6 cells or c.p.m./10 oocytes. Cells or oocytes injected with ABPL4 mRNA (25 ng/oocyte) were incubated with [³²P]MuIFN-γ with (+) or without (−) unlabelled competitor MuIFN-γ as described in *Protocols 2* and *3*. The results of a typical experiment are given for mouse ABPL4 and human HL60 cells which were used as positive and negative controls, respectively, for the binding reaction. Values for a typical experiment with non-injected and buffer-injected oocytes are also shown. All values for the oocytes represent the average of duplicate or triplicate determinations. The ratio (−/+) represents the c.p.m. bound in the absence to that in the presence of unlabelled competitor MuIFN-γ. The data were taken from Kumar *et al.* (34).

3.5 Covalent cross-linking of [^{32}P]muIFN-γ to cells and oocytes

∞Protocol 4

Covalent cross-linking of [^{32}P]muIFN-γ to cells and oocytes

Equipment and reagents

- Dulbecco's modified Eagle's medium (DMEM) supplemented with 10% bovine serum (Gibco, Sigma)
- [^{32}P]muIFN-γ (see Section 2)
- Unlabelled muIFN-γ
- RNA (prepared as in 34, 35)
- 0.5 × Leibovitz's L-15 medium
- Rotary shaker
- PBS
- DSS (see Section 2)

- 1 M Tris–HCl pH 7.5
- PBS containing 0.5% (v/v) Triton X-100 and 5 mM EDTA
- Eppendorf microcentrifuge with horizontal rotor (model 5415C, Brickman Inc).
- SDS-PAGE equipment and reagents
- Autoradiography equipment Kodak Biomax MR Film and reagents
- Intensifying screen

Method

1 Resuspend the cultured cells in medium at a density of 5 × 10^6/ml. Add about 10^6 c.p.m. of [^{32}P]muIFN-γ, with or without 1 µg of unlabelled muIFN-γ as competitor, to 0.3 ml of the cell suspension. For cross-linking to the receptor expressed in oocytes, inject with RNA, and, after about 48 h, place 50 oocytes in 800 µl of 0.5 × Leibovitz's L-15 medium containing about 6 × 10^5 c.p.m. of [^{32}P]muIFN-γ, with or without 0.6 µg of unlabelled muIFN-γ.

2 Allow binding to proceed for 90 min at room temperature. Gently resuspend cells every 15 min; incubate oocytes on a rotary shaker.

3 Centrifuge cells at 15 000 g for 20 sec in a microcentrifuge, wash twice with 1 ml of cold PBS, and finally resuspend in 500 µl of PBS. Wash the oocytes three times with cold 0.5 × L-15 medium, three times with cold PBS, and finally resuspend in 800 µl of PBS.

4 Covalently cross-link by the addition of DSS to a final concentration of 0.5 mM (5 µl of 50 mM DSS per 500 µl of PBS).

5 After 20 min on ice, quench the reactions for 5 min on ice by adding Tris–HCl pH 7.5, to a final concentration of 20 mM (10 µl of 1 M Tris–HCl, pH 7.5, per 500 µl of PBS).

6 Pellet the cells and extract with 120 µl of PBS containing 0.5% (v/v) Triton X-100 and 5 mM EDTA. Allow the oocytes to settle, carefully remove the supernatant, and extract the oocytes with 240 µl of the Triton X-100 solution.

7 After 20 min on ice, sediment insoluble material at 15 000 g for 10 min at 4 °C.

8 Analyse the supernatants by SDS-PAGE (39). Dry the gel and subject it to autoradiography with an intensifying screen.

The cross-linked [^{32}P]muIFN-γ–receptor complex (see *Figure 3*) is seen as a broad band of about 105–115 kDa. The monomeric and dimeric forms of [^{32}P]muIFN-γ can also be seen if a gradient gel system is used. It is noteworthy that the IFN-γ receptors are heavily glycosylated, with about 25–30% of the mass of the receptor consisting of carbohydrate (40, 41). The cross-linking data presented show that the receptor protein synthesized in oocytes is glycosylated to an extent similar to that seen on intact cells. However, it should also be noted that the glycosylation process in different cell types may vary somewhat and this could give rise to slightly different banding patterns when the cross-linked ligand–receptor complex is analysed on gels.

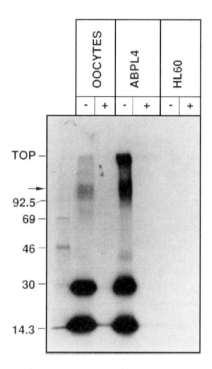

Figure 3 Covalent cross-linking of [^{32}P]muIFN-γ to receptors on cells and oocytes. [^{32}P]muIFN-γ was cross-linked to murine ABPL4 cells (suspension cells), human HL60 cells (suspension cells), or *X. laevis* oocytes injected with mRNA (25 ng/oocyte) from ABPL4 cells as described in *Protocol 4*. The detergent extracts were analysed on a 7.5–15% linear SDS-polyacrylamide gel. The binding reactions were carried out in the absence (–) or presence (+) of excess unlabelled muIFN-γ. Positions of ^{14}C-labelled molecular weight markers are indicated on the far left side of the figure. Molecular masses are expressed in kDa. The position of the broad 105–115 kDa [^{32}P]muIFN-γ·receptor complex is marked with an arrow. Since the wells for the oocyte (–) and ABPL4 cell (–) contained about the same c.p.m., it is evident that the cross-linking reaction is more efficient on cells than on oocytes. Human HL60 cells served as a negative control. The figure is reproduced from Kumar *et al.* (34).

4 Cloning the muIFN-γ receptor (muIFN-γR)

4.1 Introduction

By examining the binding and cross-linking patterns of ^{32}P-labelled hu- and muIFN-γ to panels of mouse–human, hamster–human, and hamster–mouse somatic cell hybrids, the hu- and muIFNGR1 (IFN-γR1) genes encoding the IFN-γR1 chains were localized to human chromosome 6q (19) and mouse chromosome 10 (15). During these studies, it was observed that the human and mouse homologues for the *myb* proto-oncogene are localized to the same chromosomes as those encoding the IFN-γR1 genes (42, 43). Since the IFN-γR1 and *myb* genes were genetically linked, we tested several mouse plasmacytoid lymphosarcoma cell lines for their ability to bind [^{32}P]muIFN-γ. These cell lines, established from BALB/c mice injected with the Abelson virus, exhibit enhanced transcription of *myb* sequences. We felt that, if the linkage between the two genes was close enough, the muIFN-γR1 gene might also experience increased transcription and, hence, increased translation of muIFN-γR1. One cell line, designated ABPL4, did express 5–10-fold more muIFN-γR1 chains than any other mouse cell we had previously tested.

The ABPL4 cells were used as a source of muIFN-γR1 mRNA to construct a cDNA library. We then proceeded to use the oocyte assay system to clone the muIFN-γR1 from the ABPL4 library by hybrid selection (44), hybrid-arrested translation (45), and direct expression of RNA transcripts prepared *in vitro* from pools of the cDNA library. The huIFN-γR1 cDNA clone (46) was used as a probe to screen the ABPL4 library and identify cDNAs for the muIFN-γR (35).

4.2 Procedures for isolating muIFN-γR cDNA

The procedures for isolating the muIFN-γR1 cDNA have already been described in detail (35) and will only be highlighted here:

(a) RNA was isolated from human Raji or mouse ABPL4 cells by a guanidine isothiocyanate extraction method (47), and polyadenylated mRNA was purified by two cycles of chromatography on oligo(dT)-cellulose (48).

(b) The cDNA encoding the huIFN-γR1 was cloned into a modified pGEM vector (35) and used to screen an ABPL4 cDNA library prepared as described in the following sections.

(c) The ABPL4 polyadenylated mRNA was fractionated on a 5–20% sucrose gradient as described in ref. 49. Fractions of 1 ml were collected and the RNA was precipitated with ethanol three times, washed, dried, and dissolved in sterile water.

(d) The RNA from each fraction was microinjected into oocytes as reported elsewhere (34), and binding with [^{32}P]muIFN-γ was examined about 48 h later as described in *Protocol 3*.

(e) The fraction exhibiting the highest receptor activity was used to prepare a cDNA library in the λGEM2 vector (Promega).

(f) Initially, 10 positive clones were identified that hybridized strongly to the huIFN-γR1 probe (plasmid phuIFN-γR8). These clones were then hybridized to DNA fragments which represented the binding, transmembrane, and cytoplasmic domains of the huIFN-γR1. Three clones that hybridized strongly to all three human probes were further characterized.

(g) Various fragments from the isolated clones were then subcloned into M13mp18 and M13mp19 vectors.

(h) Both strands of the largest cDNA (λmuIFN-γR4) were sequenced by the dideoxy

HOMOLOGY BETWEEN MOUSE AND HUMAN INTERFERON GAMMA RECEPTORS

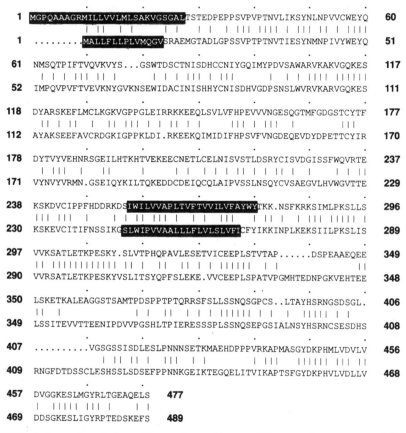

Figure 4 Comparison between the murine and human IFN-γ receptor sequences at the protein level. The first and second sequences represent the murine and human amino-acid sequences, respectively. There is 49.5% homology in the extracellular domain (murine residues 28–254), 52% homology in the transmembrane domain (residues 255–277), and 54% homology in the cytoplasmic domain (residues 278–477) between the murine and human proteins. The residues enclosed within the darkened boxes at the beginning and middle of the sequences represent the putative signal peptide and transmembrane domains, respectively. The figure is reproduced from Kumar et al. (34).

chain-termination method with Sequenase 2.0. *Figure 4* shows a comparison between the mu- and huIFN-γR1 sequences at the protein level.

(i) The insert of λmuIFN-γR4 was subcloned into vector pCK1, a pGEM4 vector modified to contain additional restriction sites (35). The resulting subclone, pmuIFN-γR36, was linearized and *in vitro* transcription was carried out with the translation cap analogue 7-methyl-G(5')ppp(5')G to increase translation of the RNA after microinjection into frog oocytes. The binding of [^{32}P]muIFN-γ to oocytes injected with sense or antisense muIFN-γR1 RNA is summarized in *Table 2*.

Table 2 Binding of [^{32}P]Mu-IFN-γ to oocytes injected with RNA transcribed *in vitro* from the Mu-IFN-γ receptor cDNA

RNA	Concentration of unlabelled Mu- or Hu-IFN-γ (competitor)	[^{32}P]Mu-IFN-γ ratio bound (c.p.m.) (−/+)	
Sense RNA	None	3180	
(pMuIFN-γR36) (0.05 ng/oocyte)	(3.1×10^{-9}M) (Mu)	647	4.9
	(3.1×10^{-8}M) (Mu)	243	13.1
	(1.6×10^{-7}M) (Mu)	176	18.0
	(3.1×10^{-7}M) (Mu)	122	26.0
	(3.4×10^{-7}M) (Mu)	3130	1.02
Sense RNA	None	18080	
(pMuIFN-γR36) (10 ng/oocyte)	(3.1×10^{-9}M) (Mu)	14550	1.2
	(3.1×10^{-8}M) (Mu)	4470	4.0
	(1.6×10^{-7}M) (Mu)	2384	7.6
	(3.1×10^{-7}M) (Mu)	1850	9.8
	(3.4×10^{-7}M) (Mu)	19580	0.92
Antisense RNA	None	106	
(pMuIFN-γR36) (10 ng/oocyte)	(3.1×10^{-7}M) (Mu)	120	0.9
None	None	115	
	(3.1×10^{-7}M) (Mu)	112	1.02

The data in the table are expressed as c.p.m./10 oocytes. Oocytes injected with RNA transcribed *in vitro* from the murine receptor cDNA clone (pMuIFN-γR36) were incubated with [^{32}P]MuIFN-γ without (−) unlabelled competitor or with increasing amounts of unlabelled (+) competitor as given in parentheses. Details are given in *Protocol 3*. All values represent the average of duplicate determinations. The ratio (−/+) represents the c.p.m. of [^{32}P]MuIFN-γ bound in the absence to that in the presence of unlabelled competitor. The data are reproduced from Kumar *et al.* (35).

Cross-linking of [^{32}P]muIFN-γ to oocytes injected with the receptor RNA indicated that the size of the major cross-linked band is about 105–115 kDa, the same as that seen by cross-linking to intact ABPL4 cells.

5 Cloning functional type I and II interferon receptors

Functional receptors for the type I (IFN-α, IFN-β, and IFN-ω) and type II (IFN-γ) interferon receptors have recently been obtained. As noted above, the IFN-γ receptor was required for binding IFN-γ to cells (19), but alone was not sufficient for activity. Another component we designated an accessory factor (AF-1) encoded on human chromosome 21 was required (12–14). AF-1 is the second chain of the IFN-γ receptor complex now designated IFN-γR2. The gene (50, 51) and cDNA (16, 52) for human IFN-γR2. were cloned. IFN-γR2 is required for signal transduction, but not for IFN-γ binding (50–52); it is a member of the class 2 cytokine receptor family of transmembrane proteins (52). A clone encoding muIFN-γR2 has also been identified (53). Thus, the IFN-γ receptor consists of at least two components: the ligand binding chain (IFN-γR1) and the second chain, IFN-γR2 (16, 52, 53).

The isolation of the gene for huIFN-γR2 was undertaken through a new technology for screening large segments of the human genome in yeast artificial chromosomes (YACs) into which mammalian selectable markers were introduced (50–52, 54, 55). As this was successful in localizing the gene for huIFN-γR2, we applied this same technology to identify a functional type I receptor. A YAC was identified that contained genes encoding the type I receptor complex (56). Furthermore, by deleting one known gene, huIFN-αR1 (57), within the YAC, we demonstrated that this gene was an essential component of the type I interferon receptor complex (58). There is at least one additional gene on this YAC clone that contributes to the binding of type I interferons and to functional activation of the receptor (56, 58; J. Cook, I. Kerr, G. Stark and S. Pestka, unpublished studies). This gene, *IFNAR2*, codes for the second chain of the type I interferon receptor, IFN-αR2c (59, 60).

6 Concluding remarks

The binding and/or cross-linking procedures outlined here may require modification when applied to other ligand–receptor systems. In our experience, we have detected no significant differences in the cross-linked bands from suspension and adherent cell lines. This indicates that the IFN-γR1 is relatively insensitive to the trypsin treatment used to release adherent cells from the tissue culture substrates. If other receptors are found to be degraded by treatment with trypsin, it may be necessary to do the binding and cross-linking procedures directly in the tissue culture plates. For binding studies, the bound radioactivity can be determined after lysing the cells with 1 M NaOH. Similarly, in a cross-linking experiment, the detergent extraction would be done in the plate and the supernatant after centrifugation would be analysed by SDS-PAGE.

The phosphorylation of interferons as described in this chapter enables proteins to be radiolabelled to very high specific activities. Although mu- and huIFN-γ have intrinsic sites for phosphorylation, genetic engineering procedures were used to extend the labelling method to other proteins such as several IFN-α species, IL-10, and monoclonal antibodies (24–33; L. Izotova, S. Kotenko, S. Saccani and S. Pestka, unpublished observations). The ^{32}P-labelled IFNs described in this chapter allowed us to set up a very sensitive expression and detection system in *X. laevis* oocytes. The assay system was also used to evaluate different tissues and cell lines for their content of receptor RNA as a prelude to cloning the IFN-γ receptor. These procedures are being used to define the activity and function of cloned receptors in heterologous cells. Furthermore, the ^{32}P-labelled ligands have been used to study binding to purified soluble receptors in a convenient and sensitive dot–blot assay (61–63).

Acknowledgements

The work presented here was supported in part by United States Public Health Services Grants RO1-CA46465 from the National Cancer Institute and RO1-AI36450, RO1-AI4369, and 5T32AI07403 from the National Institute of Allergy and Infectious Diseases. A special award from the Milstein Family Foundation provided additional support for a variety of efforts in this project.

References

1. Pestka, S. and Baron, S. (1981). In *Methods in enzymology*, Vol. 78 (ed. S. Pestka), p. 3. Academic Press, London.
2. Stewart, W. E., II (1979). *The interferon system*. Springer-Verlag, New York.
3. Pestka, S. (1983). *Arch. Biochem. Biophys.*, **221**, 1.
4. Langer, J. A. and Pestka, S. (1984). *J. Invest. Dermatol.*, **83**, 128s.
5. Pestka, S., Langer, J. A., Zoon, K. C., and Samuel, C. E. (1987). In *Annual review biochemistry*, Vol. 56 (ed. C. E. Richardson, P. D. Boyer, I. B. David, and A. Meister), p. 727. Annual Reviews, Palo Alto, CA.
6. Baron, S., Dianzani, F., Stanton, G. J., and Fleischmann, W. R., Jr. (ed.) (1987). *The interferon system: a current review to 1987*. University of Texas Press, Austin, TX.
7. Hauptmann, R. and Swetly, P. (1985). *Nucleic Acids Res.*, **13**, 4739.
8. Feinstein, S. I., Mory, Y., Chernajovsky, Y., Maroteaux, L., Nir, U., Lavie, V., and Revel, M. (1985). *Mol. Cell. Biol.*, **5**, 510.
9. Capon, D. J., Shepard, H. M., and Goeddel, D. V. (1985). *Mol. Cell. Biol.*, **5**, 768.
10. Pestka, S. (1997). *Seminars in Oncology*, **24**, S9–4.
11. Pestka, S. (1997). *Seminars in Oncology*, **24**, S9–18.
12. Jung, V., Rashidbaigi, A., Jones, C., Tischfield, J. A., Shows, T. B., and Pestka, S. (1987). *Proc. Natl. Acad. Sci. USA*, **84**, 4151.
13. Jung, V., Jones, C., Rashidbaigi, A., Geyer, D. D., Morse, H. G., Wright, R. B., and Pestka, S. (1988). *Somatic Cell Mol. Genet.*, **14**, 583.
14. Jung, V., Jones, C., Kumar, C. S., Stefanos, S., O'Connell, S., and Pestka, S. (1990). *J. Biol. Chem.*, **265**, 1827.
15. Mariano, T. M., Kozak, C. A., Langer, J. A., and Pestka, S. (1987). *J. Biol. Chem.*, **262**, 58.
16. Pestka, S., Kotenko, S. V., Muthukumaran, G., Izotova, L. S., Cook, J. R., and Garotta, G. (1997). *Cytokine Growth Factor Rev.*, **8**, 189.
17. Flores, I., Mariano, T. M., and Pestka, S. (1991). *J. Biol. Chem.*, **266**, 19875.
18. Rashidbaigi, A., Kung, H.-F., and Pestka, S. (1985). *J. Biol. Chem.*, **260**, 8514.
19. Rashidbaigi, A., Langer, J. A., Jung, V., Jones, C., Morse, H. G., Tischfield, J. A., Trill, J. J., Kung, H.-F., and Pestka, S. (1986). *Proc. Natl. Acad. Sci. USA*, **83**, 384.
20. Kung, H.-F. and Bekesi, E. (1986). In *Methods in enzymology*, Vol. 119 (ed. S. Pestka), p. 296. Academic Press, London.
21. Langer, J. A., Rashidbaigi, A., and Pestka, S. (1986). *J. Biol. Chem.*, **261**, 9801.
22. Fields, R., Mariano, T. M., Stein, S., and Pestka, S. (1988). *J. Interferon Res.*, **8**, 549.

23. Arakawa, T., Parker, C. G., and Lai, P.-H. (1986). *Biochem. Biophys. Res. Commun.*, **136**, 679.

24. Li, B.-L., Langer, J. A., Schwartz, B., and Pestka, S. (1989). *Proc. Natl. Acad. Sci. USA*, **86**, 558.

25. Zhao, X.-X., Li, B.-L., Langer, J. A., Van Riper, G., and Pestka, S. (1989). *Anal. Biochem.*, **178**, 342.

26. Wang, P., Izotova, L., Mariano, T. M., Donnelly, R. J., and Pestka, S. (1994). *J. Interferon Res.* **14**, 41.

27. Lin, L., Daugherty, B., Schlom, J., and Pestka, S. (1996). *Cancer Res.*, **56**, 4250.

28. Lin, L., Gillies, S. D., Lan, Y., Izotova, L., Wu, W., Schlom, J., and Pestka, S. (1998). *Int. J. Oncol.*, **13**, 115.

29. Lin, L., Gillies, S. D., Schlom, J., and Pestka, S. (1998). *Int. J. Oncol.*, **13**, 725.

30. Lin, L., Gillies, S. D., Schlom, J., and Pestka, S. (1998). *Anticancer Res.*, **18**, 3971.

31. Lin, L., Gillies, S. D., Schlom, J., and Pestka, S. (1999). *Protein Expr. Purif.*, **15**, 83.

32. Pestka, S., Lin, L., Wu, W., and Izotova, L. (2000). In *Methods in enzymology* (ed. J. N. Abelson, S. D. Emr and J. Thorner). Academic Press, London. (In press).

33. Pestka, S., Lin, L., Wu, W., and Izotova, L. (1999). *Protein Expr. Purif.*, **17**, 203.

34. Kumar, C. S., Mariano, T. M., Noe, M., Deshpande, A. K., Rose, P. M., and Pestka, S. (1988). *J. Biol. Chem.*, **263**, 13493.

35. Kumar, C. S., Muthukumaran, G., Frost, L. J., Noe, M., Ahn, Y. H., Mariano, T. M., and Pestka, S. (1989). *J. Biol. Chem.*, **264**, 17939.

36. Kung, H.-F., Pan, Y.-C. E., Moschera, J., Tsai, K., Bekesi, E., Chang, M., Sugino, H., and Honda, S. (1986). In *Methods in enzymology*, Vol. 119 (ed. S. Pestka), p. 204. Academic Press, London.

37. Familletti, P. C., Rubinstein, S., and Pestka, S. (1981). In *Methods in enzymology*, Vol. 78 (ed. S. Pestka), p. 387. Academic Press, London.

38. Scatchard, G. (1949). *Ann. NY Acad. Sci.*, **51**, 660.

39. Laemmli, U. S. (1970). *Nature*, **227**, 680.

40. Pestka, S., Rashidbaigi, A., Langer, J. A., Mariano, T. M., Kung, H.-F., Jones, C., and Tischfield, J. A. (1986). In *UCLA Symposium on Molecular and Cellular Biology, New Series*, Vol. 50 (ed. R. M. Friedman, T. Merigan, and T. Sreevalsan), p. 259. Alan R. Liss, New York.

41. Calderon, J., Sheehan, K. C. F., Chance, C., Thomas, M. L., and Schreiber, R. D. (1988). *Proc. Natl. Acad. Sci. USA*, **85**, 4837.

42. Dalla-Favera, R., Franchini, G. F., Martinotti, S., Wong-Staal, F., Gallo, R. C., and Croce, C. M. (1982). *Proc. Natl. Acad. Sci. USA*, **79**, 4714.

43. Sakaguchi, A. Y., Lalley, P. A., Zabel, B. U., Ellis, R., Scolnick, E., and Naylor, S. L. (1984). *Cytogenet. Cell Genet.*, **37**, 573.

44. McCandliss, R., Sloma, A., and Pestka, S. (1981). In *Methods in enzymology*, Vol. 79 (ed. S. Pestka), p. 618. Academic Press, London.

45. Paterson, B. M., Roberts, B. E., and Kuff, E. L. (1977). *Proc. Natl. Acad. Sci. USA*, **74**, 4370.

46. Aguet, M., Dembric, Z., and Merlin, G. (1988). *Cell*, **55**, 273.

47. Chirgwin, J. M., Przybla, A. A., MacDonald, R. J., and Rutter, W. J. (1979). *Biochemistry*, **18**, 5294.

48. Aviv, H. and Leder, P. (1972). *Proc. Natl. Acad. Sci. USA*, **69**, 1408.

49. McCandliss, R., Sloma, A., and Pestka, S. (1981). In *Methods in enzymology*, Vol. 79 (ed. S. Pestka), p. 51. Academic Press, London.

50. Soh, J., Donnelly, R. J., Mariano, T. M., Cook, J. R., Schwartz, B., and Pestka, S. (1993). *Proc. Natl. Acad. Sci. USA*, **90**, 8737.

51. Cook, J. R., Emanuel, S. L., Donnelly, R. J., Soh, J., Mariano, T. M., Schwartz, B., Rhee, S., and Pestka, S. (1994). *J. Biol. Chem.*, **269**, 7013.

52. Soh, J., Donnelly, R. J., Kotenko, S., Mariano, T. M., Cook, J. R., Wang, N., Emanuel, S. L., Schwartz, B., Miki, T., and Pestka, S. (1994)., *Cell*, **76**, 793.

53. Hemmi, S., Böhni, R., Stark, G., Di Marco, F., and Aguet, M. (1994). *Cell*, **76**, 803.

54. Soh, J., Mariano, T. M., Bradshaw, G., Donnelly, R. J., and Pestka, S. (1994). *DNA Cell Biol.*, **13**, 301.

55. Cook, J. R., Emanuel, S. L., and Pestka, S. (1993). *Gene Anal.: Techniques Applic.*, **10**, 109.

56. Soh, J., Mariano, T. M., Lim, J-K., Izotova, L., Mirochnitchenko, O., Schwartz, B., Langer, J., and Pestka, S. (1994). *J. Biol. Chem.*, **269**, 18102.

57. Uzé, G., Lutfalla, G., and Gresser, I. (1990). *Cell*, **60**, 225.

58. Cleary, C. M., Donnelly, R. J., Soh, J., Mariano, T. M., and Pestka, S. (1994). *J. Biol. Chem.*, **269**, 18747.

59. Domanski, P., Witte, M., Kellum, M., Rubinstein, M., Hackett, R., Pitha, P., and Colamonici, O. R. (1995). *J. Biol. Chem.*, **270**, 21606.

60. Lutfalla, G., Gardiner, K., and Uzé, G. (1993). *Genomics*, **16**, 366.

61. Stefanos, S., Ahn, Y. H., and Pestka, S. (1989). *J. Interferon Res.*, **9**, 719.

62. Stefanos, S. and Pestka, S. (1990). *J. Biol. Reg.. Homeost. Agents*, **4**, 57.

63. Puvanakrishnan, R. and Langer, J. A. (1990). *J. Interferon Res.*, **10**, 299.

Chapter 6
Cytokine signalling

B. M. J. Foxwell, A. E. Hunt, F. V. Lali, and C. Smith
Kennedy Institute of Rheumatology, 1 Aspenlea Road, Hammersmith, London, W6 8LH, UK

1 Introduction

The study of cytokines as a discrete topic has existed for approximately two decades. Initial studies focused on the identification of new cytokines, a process that is still continuing, with approximately 150 having been identified to date. The subsequent characterization of cytokine receptors led to the identification of several supramolecular cytokine receptor families (see *Figure 1*). As a consequence of these studies the complex nature of cytokine receptor structure also became apparent, with receptors often composed of multiple components. These observations gave an early indication of the complexity of cytokine signalling mechanisms. A general overview of cytokine and cytokine receptor structures can be found in the *Cytokine facts book* (1).

The mechanisms by which intracellular signalling is initiated upon binding of a cytokine to its receptor are largely dependent on the nature of the receptor. A number of cytokines of the growth-factor type (e.g. epidermal growth factor, EGF or platelet-derived growth factor, PDGF) have receptors that contain tyrosine kinase domains within the intracellular portions of the receptor, whilst transforming growth-factor β (TGF-β) receptors have integral intracellular serine/threonine kinase domains (see *Figure 1*). However, the majority of cytokine receptors do not have integral kinase or other enzymatic activities. Initial signals from these receptor families are often produced by receptor-associated cytosolic tyrosine kinases of the Jak, Src, Tec, or Syk families. A distinct signalling mechanism is employed by chemokines. These chemokines use seven transmembrane, serpentine-like receptors that are associated with trimeric (α, β, γ) G proteins (see *Figure 1*). The tumour necrosis factor (TNF) and interleukin-1 (IL-1) receptor families appear to rely on the recruitment of a number of intracellular 'adaptor-like' molecules (e.g. TRAFs, TRADD, Myd88) to initiate signalling. Following oligomerization of the receptor, these molecules associate with the receptor and recruit a number of other active species, including serine/threonine kinases.

After these initial events, signal transduction involves a large number of different mediators. These include small/monomeric G proteins (e.g. Ras), adaptor molecules (e.g. Shc, Grb 2, Crk), lipid kinases (e.g. phosphatidylinositol 3'-kinase),

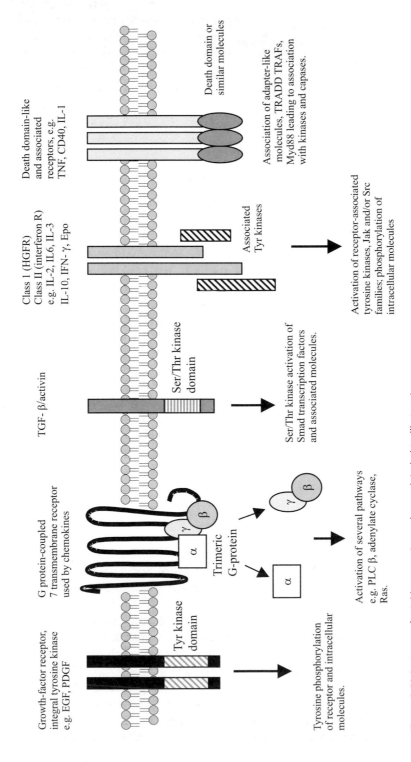

Figure 1 Major types of cytokine receptor and associated signalling systems.

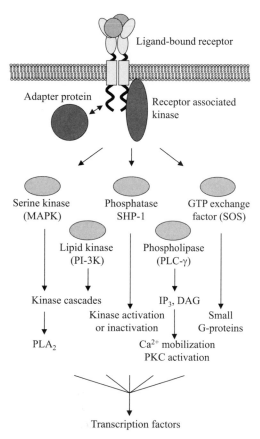

Figure 2 Generalized illustration of intracellular signalling mechanisms. Schematic representation of the signal transduction events linking receptor activation to transcription factor modulation.

serine/threonine kinase cascades (e.g. the mitogen-activated protein kinases), phospholipases, tyrosine phosphatases (e.g. SHP-1), serine and threonine phosphatases, calcium, and cAMP (*Figure 2*). Ultimately, activation of these signalling pathways results in the modulation of the activity of a variety of transcription factors leading to changes in cellular gene expression. Other important outcomes of signalling pathways include changes in the cytoskeleton or in cell motility.

It would take a large book, rather than a short chapter, to cover the whole gamut of signalling techniques that are available to, and required of, the researcher who wishes to venture into this area. By necessity, the scope of this chapter has had to be restricted. The study of cell signalling has been made a great deal easier over the last few years by the proliferation of specialist companies that provide antibodies and other reagents with which to study cell signalling. Such companies can also be a source of protocols for techniques. However, the novice should be warned that all reagents must be thoroughly checked by well-controlled experiments before any conclusions can be drawn from data, as our experience has been that many reagents do not always conform to advertised specifications.

This chapter will focus on a few of the major techniques that might be employed in investigating the signal transduction mechanisms utilized by a novel cytokine. We will not address basic biochemical techniques like immunoprecipitation and SDS-PAGE, or molecular biology techniques for making reagents, unless they impinge directly on the methods discussed below. Instead, we refer readers to the many excellent laboratory manuals for such techniques (2, 3).

2 Use of specific inhibitors to study cytokine signalling pathways

A vast array of inhibitors of enzymes involved in signal transduction are available from specialist companies (e.g. Calbiochem and Alexis), and these can provide a useful preliminary approach to signalling studies. Inhibitors of specific signalling events can be used to establish which signalling pathways are likely to be involved in mediating a particular cytokine-induced cellular effect. However, extreme caution should be exercised when interpreting data obtained using this approach. The terms 'specific' or 'highly specific' are adjectives that are often used to describe inhibitors, sometimes with little justification. Normally specificity means that *other targets have not been identified rather than the fact that other targets do not exist!* It has been our general experience that the perceived specificity of an inhibitor tends to drop with both the length of time the inhibitor has been in use and the number of different systems in which it has been applied. Another important consideration that is often overlooked in such studies is the toxicity of the inhibitor, which can vary between cell lines and types. Some assay of cell viability must therefore be included in any inhibitor study. However, despite these limitations, if data are backed up by subsequent studies that analyse signalling pathways directly (such as those discussed in this chapter), studies using inhibitors can provide useful insights and are often a good starting point for an investigation of cytokine signal-transduction mechanisms.

3 Nature and preparation of cells for signalling studies

The ability to detect the activation of specific signalling pathways in response to a stimulus is dependent upon the cells being in a resting state before the stimulus is given. The researcher should consider what cell type should be used for the study: whether they are adherent or grow as a suspension; are fully transformed, cytokine-dependent, cell lines, or primary tissue. All of which present different problems and may necessitate different approaches. Moreover, one should be aware that the signalling pathways used by a given cytokine may vary between cell types and that studies carried out using transformed cell lines may not always yield the same results as those using primary tissue.

3.1 General considerations

A number of general factors should be taken into account when preparing cells for signalling studies and, in particular, care should be taken to avoid any inadvertent activation of the cells. Some of these considerations are addressed below:

- Adherent cells should be plated out in the tissue-culture dishes/flasks in which they are to be stimulated, 24 hours prior to stimulation. The density of suspension cells should be adjusted to no more than $1-2 \times 10^6$ per ml (and sometimes less) depending upon the signalling pathway in question. The cells should be transferred to the vessel in which they are to be stimulated at least 1 hour before the start of the experiment, and allowed to equilibrate under the appropriate conditions.

- Cytokine-dependent cell lines should be washed free of any cytokine supplement before stimulation, usually at least 16–20 hours before.

- When studying signalling pathways that are strongly activated by growth factors, it may be necessary to serum-starve the cells to reduce the basal activation levels of the pathway. However, it should be noted that serum-starvation of cells, especially when prolonged, may also lead to the activation of some signalling pathways, examples of which include the c-Jun N-terminal kinase (JNK) and p38 mitogen-activated protein kinase (MAPK) pathways.

- Bacterial lipopolysaccharide (LPS) is a strong stimulus of many signalling pathways. Therefore when studying LPS-responsive cell types it is important to ensure that cell-culture reagents are endotoxin-free.

3.2 Cell lysis

The majority of signalling studies require cells to be lysed at some stage during the study. In most cases the lysis buffer will contain a detergent and a number of other supplements. In the main these will be enzyme inhibitors, normally of phosphatases and proteases, but these will obviously vary with the type of study. A key consideration is that once cell lysis occurs all reagents should be *ice-cold*. Also, since signalling studies often require comparisons to be made between cells in different states of activation, it is important to ensure that identical amounts of cellular material are used in each analysis: it is often necessary to perform assays of the cell-lysate protein concentration to check parity between samples. Basic methods for cell lysis and protein assays are given in *Protocols 1* and *2*, respectively.

Protocol 1

Preparation of cell lysates

Reagents

- Appropriate cell stimulator
- PBS
- Lysis buffer: 1% (v/v) Triton X-100, 10% (v/v) glycerol, 20 mM Hepes pH 7.4, 200 mM NaCl, 2 mM EGTA, 25 mM β-glycerophosphate. (Store at 4 °C.)
- 1 mM dithiothreitol. (Add to lysis buffer immediately prior to use.)
- Protease/phosphatase inhibitors. (Add to lysis buffer immediately prior to use):
  - (a) 1 mM phenylmethylsulfonylfluoride (Stock solution at 100 mM in isopropanol. Store at −20 °C.)
  - (b) 1 mM sodium orthovanadate (Stock solution at 100 mM in H_2O. Adjust to pH 10.0 and boil until solution is colourless. Store at 4 °C.)
  - (c) 10 mM sodium fluoride (Stock solution at 1 M in H_2O. Store at room temperature.)
  - (d) 10 µg/ml aprotinin (Stock solution at 2.5 mg/ml supplied by Sigma. Store at 4 °C.)
  - (e) 10 µg/ml leupeptin (Stock solution at 10 mM in H_2O. Store at −20 °C.)
  - (f) 10 µg/ml pepstatin (Stock solution at 10 mM in ethanol. Store at 4 °C for no more than 1 week.)

Method

1 If required, wash the cells to remove serum/growth factors/cytokines from the culture medium several hours before stimulation.

2 Stimulate the cells and then harvest after an appropriate length of time.

3 *For adherent cells stimulated in 6 or 10 cm culture* dishes:[a, b]
   - (a) Remove the culture medium from the cell monolayer and wash once with 1 ml of ice-cold PBS.
   - (b) Harvest the cells by scraping into 1 ml of ice-cold PBS.
   - (c) Transfer the cell suspension to 1.5 ml tubes and centrifuge at 10 000 g, 4 °C for 10 sec.
   - (d) Aspirate the supernatant and retain the cell pellet.

   For suspension cells stimulated in 1.5 ml tubes:
   - (a) Centrifuge at 10 000 g, 4 °C for 10 sec.
   - (b) Aspirate the culture medium and resuspend the cells in ice-cold PBS.
   - (c) Repeat the centrifugation step and aspirate the resulting supernatant; retain the cell pellet.

4 Resuspend the cell pellet in an appropriate volume of ice-cold lysis buffer containing protease and phosphatase inhibitors (100 µl of lysis buffer per 1×10^6 cells) and leave on ice for 10–20 min.

5 Centrifuge the samples at 10 000 g, 4 °C for 10 min to pellet cellular debris.

Protocol 1 continued

6   Transfer the supernatants to fresh tubes and retain on ice.

Notes and variations

[a] Cell numbers can easily be scaled up by plating adherent cells in larger tissue-culture flasks and by using larger centrifuge tubes for suspension cells. A correspondingly larger volume of PBS will be required to cover the adherent cell monolayer and cells will have to be harvested in 25 ml or 50 ml tubes by centrifugation in a normal benchtop centrifuge (2000 g, 3 min, 4°C).

[b] Adherent cells can also be lysed *in situ*, without the need for scraping the cells first. After cell stimulation, wash the monolayer once and add an appropriate volume of lysis buffer directly on to the cells. Leave the culture plate on ice for 10 min and then transfer the lysate into a 1.5 ml tube and centrifuge as described in step 4.

Protocol 2

Determination of cell-lysate protein concentration by the Bradford assay

Reagents

- Bradford reagent: 0.1% (w/v) Brilliant Blue G-250, 5% (v/v) methanol, orthophosphoric acid
- Lysis buffer containing dithiothreitol (see *Protocol 1*)

Method

1   Dilute 5 μl of each sample 1:50 in H_2O (dilution factor will have to be varied depending on the cell type and the density at which cells were lysed). Prepare a blank sample consisting of 5 μl lysis buffer diluted in H_2O to the same extent as the samples.

2   Mix 100 μl of each diluted sample with 900 μl of the Bradford reagent.

3   Use a spectrophotometer to measure the absorbance of each sample at 595 nm relative to the blank.

4   Determine the protein concentration of each sample by comparison against a standard curve constructed using the absorbance readings obtained for solutions of known concentrations of BSA.

4 Tyrosine phosphorylation and tyrosine kinases

Tyrosine phosphorylation is often the earliest detectable event in cytokine signal transduction and is therefore an obvious starting point for the investigation of cytokine signalling mechanisms. Even with receptors where there is no known integral or associated receptor, tyrosine kinase activity it may still be worth investigating this parameter.

4.1 Whole-cell phosphotyrosine Western blotting

The simplest approach to the investigation of cellular tyrosine kinase activity is to assess the extent of tyrosine phosphorylation of cellular proteins by Western blotting of whole-cell lysates. Several specific phosphotyrosine antibodies have been available commercially for a number of years, the most widely used being 4G10 (supplied by UBI). Another commonly used monoclonal antibody is PY20 (Transduction Labs) and we have obtained clear data using both antibodies. The different peptides used to generate phosphotyrosine antibodies result in some differences in their antigen specificities, and so one monoclonal antibody may not detect all tyrosine-phosphorylated species within a cell. It is therefore worthwhile considering using a cocktail of more than one antibody for Western blotting—we have often combined 4G10 with PY20.

Protocol 3

Phosphotyrosine Western blotting

Equipment and reagents

- Appropriate cell stimulator
- Lysis buffer: 0.5% (v/v) Nonidet P-40, 50 mM Tris–HCl pH 8.0, 10% (v/v) glycerol, 0.1 mM EDTA, 150 mM NaCl, 1 mM sodium orthovanadate, 10 mM sodium fluoride, 1 mM DTT, 1 mM PMSF, 10 μg/ml each of aprotinin, leupeptin, and pepstatin (see example in ref. 4) (See *Protocol 1* for preparation of inhibitors.)
- 1 × SDS-PAGE sample buffer: 62.5 mM Tris–HCl pH 6.5, 10% (v/v) glycerol, 2% (w/v) SDS, 0.00125% (w/v) Bromophenol blue, 1.4 M 2-mercaptoethanol (2-ME). (2 ×, 4 ×, and 5 × sample buffer stocks are often required and should be stored at room temperature.)
- Membrane blocking buffer (TBS): 10 mM Tris–HCl pH 7.4, 75 mM NaCl, 1 mM EDTA, 0.1% (v/v) Tween-20 5% (w/v) BSA. (This should be filter-sterilized and contain 0.1 mM sodium orthovanadate.)
- Western blotting equipment
- Rocking platform
- Membrane washing buffer: TBS containing 0.1% (v/v) Tween-20.

- Primary antibody: antiphosphotyrosine antibody(s) diluted (1:1000[a]) in TBS–Tween-20 containing 1% (w/v) BSA and 0.1% (w/v) sodium azide as a preservative. (This solution can be used several times, store at 4 °C.)
- Secondary antibody: antimurine polyclonal antibody conjugated to horseradish peroxidase (Amersham), diluted (1:5000[a]) in TBS–Tween-20 with 1% (w/v) BSA. (Make up fresh and do not add sodium azide.)
- Membranes: nitrocellulose (Schleicher and Schuell), PVDF (Millipore)
- Enhanced chemiluminescence (ECL) kit (Amersham or Pierce)
- Blot stripping buffer: 100 mM 2-ME, 2% (w/v) SDS, 62.5 mM Tris–HCl pH 6.7
- Pre-stained molecular weight markers (Amersham, Bio-Rad, or Sigma)
- SDS-PAGE equipment and reagents for 7.5% acrylamide gel
- Plastic wrap (e.g. Saranwrap®)
- Photographic film (e.g. Hyperfilm-ECL, Amersham)

Protocol 3 continued

Method

1   Stimulate cells for a short period (1–5 min), although a kinetic study over a longer time scale is preferable. Harvest cells rapidly, as described in *Protocol 1*.

2   Resuspend the cells in ice-cold lysis buffer to achieve a final density of 10^7–10^8 cells/ml and leave for 20 min on ice.[b] Centrifuge at 10 000 g, 4 °C for 10 min to remove nuclei and cellular debris.

3   Transfer the supernatants to fresh tubes and determine the protein concentration of each sample by the Bradford assay (see *Protocol 2*).

4   Add 0.25 volumes of 5 × SDS-PAGE gel sample buffer to each sample and heat at 100 °C for 3 min.

5   Resolve the samples on an SDS-PAGE gel. The percentage of acrylamide in the gel will vary depending upon the molecular weights of the proteins being analysed, typically 7.5–10% (w/v) gels are used.

6   Transfer the proteins from the gel on to a nitrocellulose or PVDF membrane, using either wet transfer or semi-dry transfer apparatus (PVDF membranes should be soaked in methanol and then equilibrated in transfer buffer prior to use). The use of pre-stained molecular weight markers when running the gel is recommended as these will help in monitoring the efficiency of protein transfer.

7   Trim the membrane if necessary and transfer to a plastic tray.[c]

8   Incubate the membrane with blocking buffer for at least 1 hour at room temperature (or overnight at 4 °C) with agitation.

9   Remove the blocking buffer and add the antiphosphotyrosine antibody solution. Incubate for 60–90 min at room temperature with constant agitation.[d]

10   Wash the membrane extensively in membrane washing buffer—three 10 min washes on a rocking platform are recommended.

11   Incubate the membrane with the secondary antibody solution for 1 h at room temperature.

12   Wash the membrane extensively as described in step 10.

13   Drain the excess wash buffer from the membrane but do not allow it to dry, especially if using PVDF membrane.

14   Incubate the membrane with ECL reagent for 1 min, drain and wrap in plastic film (e.g. Saranwrap®)

15   Expose the membrane to photographic film.

Notes and variations

[a] The optimal concentrations of the primary and secondary antibodies used in the blotting technique may vary and should be determined empirically by experimentation.

[b] An alternative method of cell lysis used to minimize tyrosine phosphatase activity involves

the addition of hot SDS-PAGE gel sample buffer (1 ×) directly on to the cell pellet. However, a problem associated with this approach is the release of DNA, which results in a dramatic increase in sample viscosity. To enable easy loading of the material on to the SDS-PAGE gel the sample should be passed through a narrow-gauge needle or rapidly freeze/thawed several times to shear the DNA.

[c] Where reagents are limiting, antibody incubation steps can be performed using smaller volumes by using a sealed plastic bag, rather than a plastic tray, to contain the membrane.

[d] Membranes can be reprobed for other proteins: incubate the membrane in stripping buffer at 60 °C for 1 h with constant agitation. Rinse the membrane with copious amounts of wash buffer before blocking and reprobing.

4.2 Investigation of tyrosine kinase activity by immunoprecipitation and phosphotyrosine Western blotting

The tyrosine kinases of the Jak family are involved in signalling from haemo-poietic growth-factor receptor (class I cytokine receptor) and interferon receptor (class II cytokine receptor) families. Phosphorylation of these kinases on tyrosine residues is intimately associated with kinase activation and can therefore be used as a marker of kinase activity. The usual method of studying Jak kinase phosphorylation is by immunoprecipitation of the protein from lysates of activated cells and subsequent phosphorotyrosine Western blotting.

Protocol 4

Analysis of Jak kinase activity by immunoprecipitation and phosphotyrosine Western blotting

Equipment and reagents

- Equipment and reagents detailed in *Protocols 1* and *3*
- Isotype-matched control antibody.
- Protein G–Sepharose (Pharmacia). (The protein-G–Sepharose should be washed three times in lysis buffer before use and used at a final suspension of 50% (v/v) in lysis buffer.)

- Anti-Jak kinase antibody (UBI or Santa Cruz)
- Wash buffer: 0.1% (v/v) IGEPAL, 50 mM Tris–HCl pH 8.0, 10% (v/v) glycerol, 0.1 mM EDTA, 150 mM sodium chloride, 1 mM sodium orthovanadate, 1 mM DTT, 1 mM PMSF, 10 μg/ml each of aprotinin, leupeptin, and pepstatin

Method

1. Lyse the cells, as described in *Protocol 3*.[a]

2. Incubate the cell lysates with 1 μg/ml control antibody and 50 μl/ml protein G–Sepharose at 4 °C for 30 min, rotating the samples constantly. Remove the protein-G–Sepharose–antibody complex (containing non-specifically associated proteins) by centrifugation at 10 000 g, 4 °C for 1 min. Transfer the supernatants to fresh tubes.

Protocol 4 continued

3   Incubate the pre-cleared lysate with 1 µg/ml anti-Jak antibody and 50 µl/ml protein-G–Sepharose[b] for 60–90 min at 4 °C, with constant rotation.

4   Pellet the Sepharose beads by centrifugation and wash the immunoprecipitates twice in lysis buffer and twice in wash buffer.

5   Add 50 µl of the wash buffer and 25 µl of 4 × SDS-PAGE gel sample buffer to the immunoprecipitates. Boil for 5 min before resolving the samples on a 7.5% acrylamide (w/v) SDS-PAGE gel.

6   Perform phosphotyrosine Western blotting as described in *Protocol 3*.[c]

Notes and variations

[a] In our experience 20–40 × 10[6] cells may be required to detect Jak phosphorylation by Western blotting. It may therefore be worthwhile increasing the cell number if initial experiments appear to fail.

[b] Protein G–Sepharose can be used to immunoprecipitate proteins using rabbit, mouse, and goat antibodies. If using an antibody of a different species origin it may be necessary to use a different matrix (e.g. protein A–agarose, Sigma) or a bridging antibody (e.g. when using rat monoclonal antibodies an anti-rat immunoglobulin antibody is required) to achieve efficient immunoprecipitation.

[c] After phosphotyrosine Western blotting it is worth stripping the blot (*Protocol 3*, note d) and reprobing it with the anti-Jak antibody used in the immunoprecipitation step, to ensure that observed variations in Jak tyrosine phosphorylation are not due to variations in the efficiency of the immunoprecipitations. The blotting procedure is as described in *Protocol 3*, although non-fat dried milk powder, rather than BSA, can be used in the blotting solutions and the inclusion of phosphatase inhibitors in the blocking solution is unnecessary.

Obviously this approach can be applied to other tyrosine kinases where changes in tyrosine phosphorylation are linked to activation and where an immunoprecipitating antibody is available. However, some kinases, for instance the Src family, are also tyrosyl-phosphorylated in the inactive state and thus may show no clear changes in net tyrosine phosphorylation upon activation. However, it may be still be possible to study the activation of such kinases by analysing tyrosine phosphorylation if antibodies are available to specific tyrosine-containing sequences whose phosphorylation is activation-linked (see ref. 5, and for a general method see Section 5.1.1). Alternatively, direct assays of kinase activity can be performed (see Section 4.3 and *Protocols 5* and *6*).

4.3 Investigation of tyrosine kinase activity by an *in vitro* kinase assay

To study the activation of the Src family tyrosine kinases, e.g. lck and fyn, we have used the immunokinase assay approach. Stimulated cells are lysed and a specific antibody is used to immunoprecipitate the kinase of interest. The immunoprecipitates are then incubated with a substrate protein or peptide, and a mixture of non-labelled and [γ-^{32}P]-labelled ATP. Kinase activity results in the catalytic transfer of a ^{32}P-labelled phosphate group to the acceptor site either on a substrate protein or the kinase itself (auto-kinase assay). Substrate proteins can subsequently be separated from unincorporated $[\gamma\text{-}^{32}\text{P}]$ATP by SDS-PAGE. Incorporation of ^{32}P-labelled phosphate into the substrate can then be assessed by autoradiography, Phosphorimager analysis, or by scintillation counting (if a peptide substrate is used). The major advantage of using this approach is that it provides a means of directly assessing the activity of specific kinases and, if phosphate incorporation is determined by Phosphorimager analysis or by scintillation counting, can provide a quantitative measure of kinase activity. However, this method relies upon the availability of both a kinase substrate and an immunoprecipitating antibody that is highly specific for the kinase of interest. The possibility of the co-immunoprecipitation of other factors along with the kinase of interest, that could affect *in vitro* kinase assays, should always be considered. We have used the following procedures (described in *Protocols 5* and *6*) to investigate the activity of the Src family kinases lck and fyn in lymphocytes (see ref. 6).

Protocol 5

Auto-kinase assay[a]

Equipment and reagents

- Serum-free medium
- Appropriate cell stimulator
- Lysis buffer: 1% (v/v) IGEAL, 25 mM Tris-HCl pH 7.5, 150 mM NaCl, 100 μM sodium orthovanadate, 10 mM sodium fluoride, 1 mM PMSF, and 10 μg/ml each of aprotinin, leupeptin, and pepstatin, 1 mM DTT (For the preparation of inhibitors refer to *Protocol 1*.)
- Bradford assay reagent (see *Protocol 2*)
- Anti-Src kinase antibody (Investigate various suppliers)
- Non-specific antibody (Investigate various suppliers)
- Protein G–Sepharose (see *Protocol 4*)
- TBS–vanadate: 20 mM Tris-HCl pH 7.5, 150 mM NaCl, 1 mM DTT, 100 μM sodium orthovanadate.

- Kinase assay buffer: 20 mM Hepes pH 7.0, 10 mM MnCl$_2$, 1 mM DTT
- ATP mix: 0.5–1 μCi/μl $[\gamma\text{-}^{32}\text{P}]$ATP (3000 Ci/mM) (Redivue, Amersham) and unlabelled ATP in kinase assay buffer
- 2 × gel sample buffer (refer to *Protocol 3*)
- SDS-PAGE equipment and reagents (see *Protocol 3*) for 8% w/v acrylamide gel
- Coomassie Blue staining solution: 0.15 % (w/v) Coomassie Brilliant Blue G, 50% (v/v) methanol, 20% (v/v) glacial acetic acid
- Destaining solution: 30% (v/v) methanol, 1% (v/v) formic acid
- Whatman 3MM paper
- Autoradiography equipment and reagents *or* equipment and reagents for Phosphorimager analysis

Method

1 Equilibrate the cells in serum-free medium using approximately 1×10^7 cells per assay (refer to *Protocol 1*).

2 Stimulate the cells appropriately and then harvest by centrifugation, as described in *Protocol 1*.

Protocol 5 continued

3 Lyse the cells in 0.5 ml ice-cold lysis buffer for 10–20 min and then remove nuclei and cellular debris by centrifugation (see *Protocol 1*). Transfer the supernatants to fresh tubes and determine the protein concentrations by the Bradford assay, as described in *Protocol 2*.

4 Pre-clear the cell lysate by incubating with 1–5 μg/ml of the non-specific isotype-matched control antibody and 50 μl protein G–Sepharose per ml of lysate for 30 min.

5 Incubate the pre-cleared cell lysates with the antikinase antibody (1:100–1:200 dilution of antiserum or 10 μg/ml purified antibody) for 1 h, on ice. Note that the immunoprecipitation antibodies should not target the active site of the kinase or otherwise interfere with kinase activity.

6 Add 20 μl/ml protein G–Sepharose and incubate for a further for 1–2 h at 4 °C with constant rotation.

7 Centrifuge the samples (10 000 g, 1 min, 4 °C) to harvest the protein-G/Sepharose/antibody/kinase complex and discard the supernatant.

8 Wash the immunoprecipitates three times in lysis buffer and once in TBS–vanadate.

9 Resuspend the immunoprecipitates in 10 μl of the kinase assay buffer and then add 10 μl of the labelled ATP mix to each sample (include a control containing un-labelled ATP mix). Incubate on ice for 10 min.

10 Centrifuge the samples briefly (10 sec) to clear the lids of the tubes and then quench the kinase reaction by adding 30 μl of 2 × gel sample buffer and heat the samples for 3 min at 100 °C.

11 Resolve the samples on an 8% (w/v) SDS-PAGE gel.

12 If the dye front has been allowed to run off the gel, carefully dispose of the radioactive buffer in the lower reservoir of the electrophoresis apparatus. If the dye front is still on the gel, cut the gel just above the dye front and dispose of the lower portion as high-level solid waste.

13 Transfer the gel into a tray containing Coomassie Blue staining solution and agitate gently at room temperature for 5–10 min.

14 Remove the Coomassie Blue staining solution and wash the gel in destaining solution with several changes of destain for at least 1 h, or until the protein bands are visible.[b]

15 Dry the gel on to Whatman 3MM paper and determine the level of [γ-^{32}P]phosphate incorporation by autoradiography or by Phosphorimager analysis.

Notes and variations

[a] An alternative approach to assaying lck using phosphopeptide affinity chromatography has been published by Seih *et al.* (see ref. 7). This method relies upon the fact that the SH2 domain of lck becomes exposed following kinase activation. Researchers wishing to avoid using radioactivity might wish to investigate this technique.

[b] Proteins can be transferred on to a nitrocellulose membrane prior to autoradiography. This method permits subsequent Western blotting to determine whether equal amounts of the kinase have been immunoprecipitated from each sample.

Protocol 6

Src kinase assays using an enolase substrate

Equipment and reagents

- Substrate: rabbit muscle enolase (ammonium sulfate precipitate, Sigma)
- Enolase buffer: 1 mM DTT, 50 mM Hepes pH 7.0
- Glycerol
- 50 mM acetic acid
- 2 × kinase assay buffer: 40 mM Hepes pH 7.0, 4 mM MnCl$_2$, 2 mM DTT (a 1 × stock will also be needed)
- 1 M Hepes pH 7.0

- ATP mix: 1 μM unlabelled ATP, 1 μCi/μl [γ-^{32}P]ATP (3000 Ci/mM) in 1 × kinase assay buffer
- 4 × SDS-PAGE gel sample buffer (see *Protocol 3*)
- SDS-PAGE equipment and reagents (see *Protocol 3*)
- Analysis equipment and reagents as for *Protocol 4*

Method

1 Add 20 μl enolase buffer per 100 μg of enolase, mix well and leave for 30–60 min on ice.

2 Add an equal volume of glycerol, mix and keep on ice if to be used on the same day. Alternatively store at −70 °C in 50 μl aliquots (enough for 25 assays).

3 Immediately prior to assay, add 50 μl of 50 mM acetic acid per 50 μl aliquot of enolase, mix and incubate for 5 min at 30 °C to denature the enolase.

4 Add 25 μl of 1 M Hepes pH 7.0 and 125 μl of 2 × kinase assay buffer to the acid-denatured enolase and keep on ice until required, but for no longer than 1 h.

5 Immunoprecipitate the Src kinase as described in *Protocol 5*. Resuspend the washed immunoprecipitates in 10 μl of the 1 × kinase assay buffer and then add 10 μl of acid-denatured enolase and 10 μl of the ATP mix (include a control containing unlabelled ATP mix). Mix the samples and incubate them for 5 min on ice, followed by 10 min at 30 °C with constant agitation.

6 Centrifuge the samples briefly (10 sec) to clear the lids of the tubes and then quench the kinase reaction by adding 15 μl of 4 × gel sample buffer. Analyse substrate phosphorylation, as described in *Protocol 4*.

5 Serine/threonine kinases

Serine/threonine kinases are generally activated later in signal transduction pathways than tyrosine kinases and are often activated by tyrosine phosphorylation. Serine/threonine kinases phosphorylate an enormous range of cytoplasmic and nuclear proteins. These kinases are often found as components of signalling kinase cascades, e.g. p38 mitogen-activated protein kinases (MAPKs) which are discussed below. Other examples of serine/threonine kinases are protein kinases A, B, and C (PKA, PKB, and PKC) and the ribosomal S6 subunit kinase p70 S6-

kinase. More recently, identified examples are the IκB kinases (IKKs) which are involved in signalling leading to the activation of the transcription factor NFκB (nuclear factor-κB). Clearly the large number of serine/threonine kinases means that this discussion of the approaches employed to assess their activity must be restricted to specific examples. However, it should be possible to adapt the techniques discussed here to investigate other serine/threonine kinases without too much difficulty.

The MAPKs are a family of serine/threonine protein kinases that are activated by a wide variety of stimuli, including cytokines. MAPKs are activated by phosphorylation of the tyrosine and threonine residues within a TXY activation motif, with phosphorylation of both residues being required for kinase activation. The MAPKs can be classified into three major groups on the basis of the identity of the intervening X residue in the TXY motif, with a number of different isoforms constituting each group. The p42/44 MAPK or extracellular signal regulated kinases (ERK2 and ERK1, respectively) have a TEY activation motif. The c-Jun N-terminal kinases (JNKs) or stress-activated protein kinases (SAPKs) have a TPY activation motif, whereas the p38 MAPKs have a TGY activation motif.

The phosphorylation of the tyrosine and threonine residues is performed by an upstream, dual-specificity MAPK kinase (MKK). Specific MKKs have been identified as being responsible for the activation of each of the three MAPKs. MKKs are themselves activated by the phosphorylation of serine residues by a MAPK kinase kinase or MKKK. Thus the activation of MAPKs involves the sequential activation of a number of serine/threonine kinases in what is termed a 'kinase cascade'.

5.1 Analysis of MAPK pathway activity

Generally, MAPKs are rapidly activated in response to cytokines, with activity reaching a maximum after approximately 20 min and returning to basal levels within 1–2 hours. However, in some cases a later period of sustained MAPK activation may be observed, so time-course studies are well worth doing. As for the tyrosine kinases, two main approaches are widely utilized to study the activation of MAPK pathways:

- detection of activating phosphorylation(s) by Western blotting; and
- direct measurement of MAPK activity by *in vitro* kinase assay.

Each of these approaches has advantages and disadvantages, which are discussed in the relevant sections.

5.1.1 Analysis of MAPK phosphorylation by Western blotting

This approach relies on antibodies raised against peptides containing the phosphorylated activation motifs of active MAPKs (i.e. phosphorylated upon both the threonine and the tyrosine residues), which are now commercially available. This approach to the study of MAPKs has several advantages:

- Only small numbers of cells are required to detect active MAPKs—this is a major advantage over the *in vitro* kinase assay approach in studies of primary cells or tissues, where cell numbers may be limiting.

- Phospho-MAPK antibodies can be highly specific.
- Western blotting is comparatively simple and no radioactive isotopes are required.

However, it should be considered that although phosphorylation of MAPKs on tyrosine and threonine has been shown to correlate with their activity, other regulatory mechanisms that do not affect these residues may also exist and thus, ideally, phosphospecific Western blotting data should be confirmed by *in vitro* kinase assay.

This approach is also applied in the study of many other serine/threonine kinases where defined phosphorylation sites have been linked to kinase activation (e.g. PKB, p70 S6 kinase, and the MKKs, the kinases immediately upstream of the MAPKs). Such phosphospecific antibodies to PKB and the MKKs are available from New England Biolabs and can be used following *Protocol 8*. In addition, phosphospecific antibodies to a number of MAPK substrates (Elk-1, c-Jun, ATF-2) can be obtained (New England Biolabs), and may provide further information on the activation of MAPK pathways. However, given the considerable overlap in the substrates utilized by specific MAPKs, such experiments may provide only limited information on the activation of specific MAPK pathways.

Protocol 7

Assessment of MAPK activation by phosphospecific Western blotting

Equipment and reagents

- Appropriate cell stimulator
- Lysis buffer: as described in *Protocol 1*
- Bradford assay reagent (see *Protocol 2*)
- Anti-phosphoMAPK antibodies (New England Biolabs)
- Other buffers and equipment (including that for SDS-PAGE and Western blotting (use a PVDF membrane)) as described in *Protocol 3*
- Secondary HRP-conjugated antibody
- 10% w/v SDS-PAGE gel
- TBS–Tween (see Membrane blocking buffer, *Protocol 3*)
- 5% (w/v) BSA in TBS–Tween
- 5% (w/v) non-fat dried milk powder in TBS–Tween
- ECL equipment and reagents (see *Protocol 3*)

Method

1 Stimulate and lyse the cells as described in *Protocol 1*.

2 Determine the protein concentration of the cell lysates by Bradford assay as described in *Protocol 2*.

3 Load equivalent amounts of lysate protein on to a 10% w/v SDS-PAGE gel.

4 After electrophoresis, transfer the proteins to a PVDF membrane as described in *Protocol 3*.

5 Block the and wash membrane as in *Protocol 3*.

6 Incubate the membrane with the phosphoMAPK antibody (diluted 1:1000 in 5% (w/v) bovine serum albumin (BSA) in TBS–Tween, see *Protocol 3*) for at least 3 h at room temperature or at 4 °C overnight.[a]

7 Remove the primary antibody solution and wash the membrane with TBS–Tween three times, for 10 min each time.

8 Incubate the membrane with the secondary HRP-conjugated antibody (diluted 1:2000 in 5% (w/v) non-fat dried milk powder in TBS-Tween for 1 h at room temperature, with agitation (for details see *Protocol 3*).

9 Wash the membrane extensively with TBS–Tween—at least four times, for 10 min each time.

10 Develop the blot with ECL as described in *Protocol 3*.[b]

Notes and variations

[a] The anti-phosphoMAPK kinase antibody solution can be used several times, store the solution at 4 °C.

[b] Following development the blot should be stripped and reprobed with antibodies to detect total MAPK, to ensure equal protein loading on the gel and to confirm that the correct kinase is being detected.

5.1.2 Analysis of MAPK activity by *in vitro* kinase assay

A wide variety of antibodies to MAPK family members are now commercially available, enabling the specific immunoprecipitation of certain MAPK isoforms. In our studies we have made use of a rabbit antiserum raised against a peptide from the C-terminal region of p38 MAPK, as well as antisera to a C-terminal peptide of full-length JNK2 and a C-terminal peptide of the short (46 kDa) form of the JNK3 isoform. In addition, affinity-purified antibodies to JNK1 (Pharmingen) and p42 MAPK (Santa Cruz Biotechnology) have been used to immunoprecipitate MAPKs.

A substrate for the kinase of interest is also required in order to carry out *in vitro* kinase assays. Bacterially expressed glutathione S-transferase (GST) or histidine $(His)_6$-tagged fusion proteins are widely used, as the presence of the GST or $(His)_6$ allows the simple purification of the substrate protein from bacterial cell lysates by affinity chromatography. Glutathione-linked Sepharose can be used to bind GST-linked proteins, these can be subsequently eluted from the Sepharose beads with free glutathione. Divalent metal (nickel or copper)-linked matrices (Ni-NTA–agarose, Qiagen, or Talon resin, Clontech) can be used to bind His_6-tagged proteins and these can be eluted from the matrix with imidazole. We have found that both GST–ATF2 (residues 9–89) and GST–c-Jun (residues 1–135) are highly efficient substrates for JNK. GST–ATF2 can also be used as a substrate for p38 MAPK, but we have found that a recombinant, His_6-tagged, full-length murine MAPKAPK-2 (MAPK activated protein kinase-2) is a far more efficient substrate for this kinase. p42/44 MAPK activity is usually assayed using myelin basic protein (MBP, Sigma) as a substrate.

Protocol 8

Analysis of MAPK activity by *in vitro* kinase assay[a]

Equipment and reagents

- Lysis buffer and inhibitors (as described in *Protocol 1*)

- Bradford assay reagent (see *Protocol 2*)

- Anti-MAPK immunoprecipitating antibody

- RIPA buffer: 20 mM Hepes pH 7.4, 20 mM β-glycerophosphate, 0.5 mM EDTA, 0.5 mM EGTA, 0.2 M NaCl, 2 mM DTT, 1% (v/v) IGEPAL 0.5% (w/v) sodium deoxycholate, 0.1% (w/v) SDS (Store at 4 °C. Add 1 mM sodium orthovanadate, 1 mM DTT immediately prior to use.)

- Kinase assay buffer: 50 mM Tris pH 7.4, 20 mM β-glycerophosphate, 20 mM $MgCl_2$ (Store at 4 °C. Add 1 mM sodium orthovanadate, 1 mM DTT immediately prior to use.)

- Protein G–Sepharose (see *Protocol 4*)

- Substrate solution: 66 μg/ml MBP for p42 MAPK; 33 μg/ml GST–ATF2 (9–109) *or* GST-Jun (1–135) for JNK; 66 μg/ml GST–ATF2 (9–109) *or* 100 μg/ml His_6-MAPKAPK-2 for p38 MAPK (All in kinase assay buffer.)

- ATP mix: 40 μM ATP (store 6 mM aliquots at −20 °C), 0.25 μCi/μl [γ-^{32}P]ATP (3000 Ci/mM) (Redivue Amersham) in kinase assay buffer

- SDS-PAGE equipment and reagents (see *Protocol 3*); 4 × gel sample buffer, 12.5% (w/v) SDS-PAGE gel

Method

1 Stimulate and lyse cells as described in *Protocol 1*. Note that the number of cells required will depend on cell type, but use approximately 5×10^6 primary human monocytes or RAW 264.7 murine macrophages for an MAPK *in vitro* kinase assay.

2 Determine the protein concentration of the cell lysates by Bradford assay as described in *Protocol 2*.

3 Transfer lysates of equivalent protein concentration (approximately 1 mg total protein) to fresh microcentrifuge tubes and adjust the volumes of the samples to 1 ml with lysis buffer.

4 Add the anti-MAPK antibody to each sample. Use a dilution of 1:100–1:200 of anti-serum: sample. For an affinity-purified antibody, use a final antibody concentration of approximately 1–5 μg/ml.

5 Incubate at 4 °C for 1 h before adding 30 μl/ml of protein G–Sepharose. Incubate the samples for a further 2 h at 4 °C, rotating the samples continuously (see *Protocol 4*).

6 Add 1 ml of ice-cold RIPA buffer to each immunoprecipitate, vortex briefly to resuspend the Sepharose beads, and microcentrifuge at maximum speed, at 4 °C for 1 min. Aspirate the supernatant from the beads and repeat the washing process twice.

7 Wash the immunoprecipitates twice with 1 ml of the kinase assay buffer, taking care to aspirate all the assay buffer after the final wash.

8 Add 20 μl of the appropriate substrate solution to each sample.

9 Add 20 μl of the ATP mix to each sample (include an unlabelled ATP control sample) and incubate on a shaker at room temperature for 20 min.

10 Centrifuge the samples briefly (10 sec) to clear the lids of the tubes. Stop the kinase reaction by adding 20 μl of the 4 × gel sample buffer (see *Protocol 4*).

11 Resolve the samples on a 12.5% (w/v) SDS-PAGE gel, Coomassie stain, and dry the gel. Analyse substrate phosphorylation as described in *Protocol 5*.

Notes and variations

[a] Sometimes a substrate peptide (rather than a protein) can be used. The radioactivity incorporated into the substrate peptide can either be analysed by blotting the kinase mixture on to P81 Whatman filter paper. P81 paper squares are then washed extensively in 0.75% (v/v) orthophosphoric acid, dried, and analysed by scintillation counting. Alternatively, substrate peptides can be resolved by appropriate SDS-PAGE (ref. 7).

6 Transcription factors

As described previously, the most commonly studied outcome of intracellular signalling is the modulation of gene transcription. In both prokaryotes and eukaryotes gene transcription is regulated by the binding of nuclear proteins, or transcription factors, to specific DNA sequences, known as promoter or enhancer elements. The transcriptional machinery of the cell comprises two parts: basal and induced. The basal transcriptional elements include constitutively active RNA polymerases and associated binding factors. These bind to the promoter site approximately 25 base pairs upstream of the transcriptional start site in the region of the TATA box. Interaction of the basal transcriptional machinery with a host of inducible transcription factors, which are the targets of signalling cascades, is required to effect significant changes in gene transcriptional activity. Differential regulation of combinations of inducible transcription factors by different cell stimuli can result in the specific activation of subsets of cellular genes.

Investigation of the inducible transcription factors regulated by a cytokine stimulus can be achieved by measuring physical changes at more than one level. For example, as a consequence of kinase cascade activation, transcription factor activities are often induced by phosphorylation, a process that can be monitored by Western blotting (see *Protocol 3*). The translocation of transcription factors between distinct subcellular compartments can also be assessed by cell fractionation and subsequent Western blotting (e.g. NFκB is translocated from the cytoplasm to the nucleus upon its release from an inhibitory factor, IκB). Finally, the interaction of inducible transcription factors with specific DNA sequences can be monitored. Two approaches to the study of transcription factor–DNA binding are outlined below.

6.1 Studying the activation of STAT transcription factors by oligoprecipitation

The binding of transcription factors to defined nucleotide sequences can be monitored with a precipitation approach using biotinylated oligonucleotides. Transcription factors that are active and bind to the biotinylated consensus oligonucleotide sequences are precipitated using streptavidin-coupled agarose beads, which can then be detected by Western blotting. This approach has been particularly useful for investigating STAT (signal transducer and activator of transcription) activation (see *Protocol 9*).

Protocol 9

Oligoprecipitation for studying the activation of STAT transcription factors

Equipment and reagents

- Streptavidin-agarose/Sepharose (4% beaded agarose, Cynagen bromide linked) (Sigma) (Wash beads three times in lysis buffer before use.)
- Antibodies for Western blotting (Various suppliers.)
- Other reagents and requirements are as described in *Protocols 3* and *4*.
- Biotinylated oligonucleotides (Double-stranded consensus oligonucleotides are 5′-biotinylated using an appropriate custom carbon-chain spacer. Such modified oligonucleotides can be obtained from specialist companies.)

Method

1 Stimulate the cells and prepare lysates as described in *Protocol 3*. STAT activation (tyrosine phosphorylation) is detectable within 5–15 min of stimulation.

2 Pre-clear the cell lysates by rotation with 50 µl/ml streptavidin–agarose for 30 min at 4 °C. Centrifuge at 10 000 *g*, at 4 °C for 1 min to pellet the agarose beads and transfer the supernatants to fresh tubes.

3 Add the biotinylated, double-stranded, consensus oligonucleotide to the supernatant (1 µg/ml lysate), mix and incubate for 30–60 min at 4 °C.

4 Add the streptavidin–agarose beads to the lysate (25 µl/ml) and incubate with constant rotation for a further 30 min at 4 °C.

5 Wash the agarose beads three times in 1 ml of lysis buffer. Resuspend the beads in 50 µl lysis buffer and add 25 µl 4 × gel sample buffer. Heat the samples to 100 °C for 3 min and resolve them on an SDS-PAGE gel (acrylamide concentration 7.5–10% (w/v)).

Protocol 9 continued

6   Transfer the proteins to nitrocellulose or PVDF membranes and perform Western blotting (as described in *Protocol 4*) using antibodies to phosphotyrosine or to specific STAT proteins. Note that antibodies to phosphorylated STATs are also available and can be used in Western blotting of whole-cell lysates.

Notes and variations

This Protocol can be combined with studies on Jak kinases. Once a lysate has been used for a Jak immunoprecipitation the same lysate can be analysed for STATs using the above Protocol. To reduce time, streptavidin–Sepharose can be included with protein-G–Sepharose into the initial pre-clearing step. In our experience STAT signals are much stronger than Jak signals and we therefore advise performing the Jak immunoprecipitation first.

6.2 Investigation of transcription factor activation by electrophoretic mobility-shift assay (EMSA)

6.2.1 General considerations

EMSA takes advantage of the fact that protein–DNA complexes migrate more slowly than free DNA in a non-denaturing polyacrylamide gel. By radiolabelling a consensus DNA fragment which specifically binds a particular transcription factor, the DNA binding activity of a particular transcription factor in a nuclear protein extract can be monitored using autoradiography or Phosphorimager analysis. This approach first requires that proteins are prepared from nuclei. Nuclei can be isolated from whole cells by detergent lysis under hypotonic conditions as described in *Protocol 10*. Lysis in the presence of low concentrations of salt causes the nuclear pores to close, preventing proteins leaching from the nucleus. The nuclei can then be pelleted by centrifugation and washed to remove residual traces of cytoplasmic material. The cytoplasmic supernatant can be retained for Western blotting to follow subcellular translocation events.

Protein extraction from the nuclei is achieved in buffers containing high salt concentrations. Hypertonic conditions cause the nuclear pores to open, allowing an efflux of non-histone-associated nuclear proteins and leaving an insoluble complex of DNA and histones within the nuclei. Most sequence-specific transcription factors can be extracted using 0.3–0.4 M NaCl. The ionic strength of the extraction buffer described in *Protocol 10*, which contains 0.5 M NaCl, can be lowered to this level if problems of non-specific DNA binding activities are encountered.

6.2.2 Radiolabelling consensus oligonucleotides

The double-stranded oligonucleotides can either be synthesized by annealing two complementary synthetic oligonucleotides (*Protocol 11*, Part A) or by purchasing them pre-annealed (Promega). The double-stranded consensus oligonucleotides can be purchased in blunt-ended form, without 5'-phosphate groups, which can then be end-labelled using T4 polynucleotide kinase in the presence of

[γ-^{32}P]ATP, as described in *Protocol 11*, Part B. The sensitivity of the EMSA assay can be enhanced by increasing the amount of radiolabel incorporated into the oligonucleotide. This is achieved by annealing oligonucleotides that generate 5'-overhanging ends to the consensus binding sequences. These overhanging regions can then be filled in using T4 DNA polymerase or Klenow with a mixture of dNTPs, including one radiolabelled dNTP. It is important to note that this procedure requires the use of [α-^{32}P]dNTP, as opposed to [γ-^{32}P]ATP, for end-labelling—use of the latter would result in the loss of the radiolabel (in the form of pyrophosphate) during the polymerase reaction.

Protocol 10

Preparation of cytosolic and nuclear protein extracts

Equipment and reagents

NB: All reagents should be ice-cold

- Hypotonic lysis buffer: 5 mM Hepes pH 7.9, 10 mM KCl, 1.5 mM $MgCl_2$

- Protease inhibitors: 10 μg/ml aprotinin, 10 μM E64, 1 μg/ml pepstatin, and 1 mM phenylmethylsulfonylfluoride (Add the protease inhibitors to both the lysis and extraction buffers immediately before use. These are normally kept as stock solutions (see *Protocol 1*).)

- 10% (v/v) IGEPAL (in water)

- PBS

- Hypertonic lysis buffer: 5 mM Hepes pH 7.9, 0.25% (v/v) IGEPAL, 25% (v/v) glycerol, 500 mM NaCl, 1.5 mM $MgCl_2$, 0.2 mM EDTA

- Rotary mixer

- Bradford assay reagent (see *Protocol 2*)

A. Isolation of nuclei

1   Stimulate the cells (typically 5–10 $\times$ 10^6 cells per condition) for the required length of time.

2   Resuspend the cell pellets in 400 μl of ice–cold hypotonic lysis buffer.

3   Add 8 μl of 10% (v/v) IGEPAL, to achieve a final concentration of 0.25% (v/v), and vortex vigorously for 10 sec. Leave on ice for 10 min.

4   Harvest the nuclei by centrifugation at 10 000 g, 4 °C for 30 sec.

5   Remove the cytosolic supernatant (retain if required and store at −70 °C) and wash the nuclear pellet once with 100 μl of ice–cold hypotonic lysis buffer.

B. Nuclear protein extraction

1   Resuspend the nuclear pellets in 50–80 μl of ice–cold hypertonic extraction buffer.

2   Extract the proteins from the nuclei by rotating the samples for 2 h at 4 °C.

3   Pellet the nuclei by centrifugation at 10 000 g, at 4 °C for 10 min.

4   Remove the supernatant and store in aliquots at −70 °C.[a]

5   Determine the protein concentration of the nuclear extracts by Bradford assay (as described in *Protocol 2*).

Notes and variations

[a] It is essential that nuclear protein extracts are stored below –70 °C otherwise protein–DNA binding activity will be lost. Protein–DNA binding activity is also reduced by freeze/thaw cycles so aliquots should be made appropriately.

Protocol 11

Radiolabelling consensus oligonucleotides

Equipment and reagents

- 20 × oligonucleotide annealing buffer: 200 mM Tris–HCl pH 7.9, 1 M NaCl, 40 mM MgCl$_2$, 20 mM EDTA
- Double–stranded consensus oligonucleotide
- 10 × T4 polynucleotide kinase buffer: 700 mM Tris–HCl pH 7.6, 100 mM MgCl$_2$, 50 mM DTT
- T4 polynucleotide kinase (10 U/μl) (New England Biolabs)
- Nuclease-free water

- [γ-^{32}P]ATP: 3000 Ci/mmol at 10 μCi/μl (Amersham)
- TE buffer: 10 mM Tris–HCl pH 8.0, 1 mM EDTA
- Sephadex Microspin G-25 spin column (Pharmacia)
- 90 °C heating block or waterbath
- 0.5 M EDTA pH 8.0
- Scintillation counter equipment and reagents

A. Oligonucleotide annealing reaction

1   Add 35 pmol of each complementary oligonucleotide to a 1.5 ml tube, together with 1 μl of the 20 × oligonucleotide annealing buffer.

2   Adjust the volume to 20 μl using nuclease-free water. Mix gently by pipetting.

3   Heat the tube and its contents to 90 °C in a heatblock or waterbath for 5 min and then **slowly** cool to room temperature by switching off the power source.

B. Labelling reaction (for blunt-ended oligonucleotides)

1   Assemble the following components in a 1.5 ml tube:

Consensus DNA probe (1.75 pmol/μl)	2 μl
10 × T$_4$ polynucleotide kinase buffer	1 μl
[γ-^{32}P]ATP (3000 Ci/mmol at 10 μCi/μl)	1 μl
T4 polynucleotide kinase (10 U/μl)	1 μl
Nuclease-free water	5 μl

Protocol 11 continued

2 Incubate the mixture at 37 °C for 10 min, then terminate the reaction by adding 1 μl of 0.5 M EDTA pH 8.0.

3 Add 39 μl of TE buffer to bring the volume up to 50 μl.

C. Removal of unincorporated radionucleotide

1 Equilibrate a Sephadex Microspin G-25 column with two 50 μl washes using TE buffer. Centrifuge at 1000 *g* for 2 min after each wash.

2 Place the column in a fresh 1.5 ml tube in which the labelled probe will be collected. Apply the 50 μl labelling reaction mixture to the column and centrifuge at 1000 *g* for 2 min.

3 Determine the activity of the radiolabelled oligonucleotide probe by scintillation counting. Typical activities achieved range between 40 000 and 100 000 c.p.m./μl.

4 Store the radiolabelled DNA consensus binding-site probe at −20 °C. The probe can be freeze/thawed many times and can be used until the activity falls below 10 000 c.p.m./μl.

Notes and variations

DNA consensus binding-site probes for established transcription factors such as AP-1, AP-2, CREB, NFκB, and Oct-1 can often be purchased directly from manufacturers, in a double-stranded form without overhangs, obviating the requirement for the oligonucleotide annealing step.

6.2.3 Electrophoretic mobility-shift assay (EMSA)

The EMSA procedure, described in *Protocol 12*, is a sensitive assay for determining the DNA binding activities of proteins to specific DNA sequences (see *Figure 3*). Ideally, the non-denaturing 5% (w/v) polyacrylamide gel should be prepared the

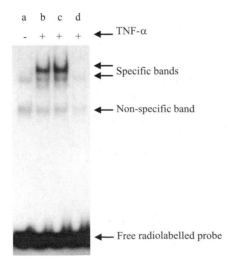

Figure 3 Electrophoretic mobility-shift assay (EMSA) from extracts prepared from HeLa cells. 5 × 10⁶ HeLa cells were treated in the presence or absence of 25 μg/ml TNF-α. Nuclear extracts were prepared as detailed in *Protocol 10*. Equal quantities of protein from each nuclear extract were incubated with the radiolabelled NFκB consensus oligonucleotide (see *Protocol 11*) and resolved on a non-denaturing 5% (w/v) polyacrylamide gel (see *Protocol 12*). The DNA binding specificity of the retarded complexes was determined by competition with no competitor (a + b), a 100-fold excess of non-labelled Oct-1 non-specific competitor DNA (c), and a 100-fold excess of non-labelled NFκB specific competitor DNA (d).

day before beginning the assay, allowing it to polymerize slowly at 4 °C (see *Protocol 12*, Part A). The DNA binding reaction (described in Part B) can be carried out at room temperature or on ice without noticeable difference. As nuclear extracts contain a large number of diverse DNA-binding proteins, Poly dI–dC is included in the binding buffer to minimize non-specific protein–labelled oligo-nucleotide interactions. *Protocol 12* has been optimized for the study of NF-κB DNA binding, therefore the level of poly dI–dC included in the binding reaction may have to be varied for other studies.

Protocol 12

EMSA/gel retardation assay

Equipment and reagents

- 5 × DNA binding buffer: 50 mM Tris–HCl pH 7.6, 20% (v/v) glycerol, 250 mM NaCl, 5 mM MgCl$_2$, 2.5 mM EDTA, 2.5 mM DTT, 0.25 mg/ml poly dI–dC (Pharmacia)
- 5% (w/v) non-denaturing polyacrylamide gel (see Part A, step 1)
- 10 × TBE stock: 0.89 M Tris–HCl pH 8.0, 0.89 M boric acid, 2 mM EDTA
- 3 litres of 0.5 × TBE buffer pH 8.3

- 10 × gel loading buffer: 250 mM Tris–HCl pH 7.5, 0.2% (w/v) Bromophenol blue, 40% (v/v) glycerol
- Nuclease-free water
- Whatman 3MM filter paper
- Autoradiography (use Hyperfilm MP (Amersham Life Sciences)) or Phosphorimager equipment and reagents

A. Preparation of the 5% (w/v) non-denaturing polyacrylamide gel

1 . To produce a non-denaturing gel (≈ 150 mm × 120 mm × 1 mm) mix the following components in a 50 ml plastic tube:

10 × TBE buffer	1.5 ml
80% (v/v) glycerol	1.0 ml
30% (w/v) acrylamide [37.5:1 *bis*]	5.0 ml
Distilled water	21.0 ml
10% (w/v) ammonium persulfate	200 μl
TEMED	20 μl

2 Pour the gel and insert the gel spacer comb immediately. Allow the gel to poly-merize over 2 h at room temperature or, preferably, overnight at 4 °C.

3 Remove the gel spacer combs carefully and wash away non-polymerized acrylamide with distilled water.

4 Pre-electrophorese the gels in 0.5 × TBE pH 8.3 for 1–2 h at 200 V or until the current drops to half its original level.

B. Nuclear protein–DNA binding reaction

1 For each experimental sample take approximately 10 μg of nuclear extract and adjust the volume to 7 μl with nuclease-free water in a 1.5 ml tube.

Protocol 12 continued

2   Set up four similar control reactions of 7 μl final volume:

(a) positive control extract;

(b) no extract (negative control);

(c) positive control extract with a 100-fold excess of non-labelled oligonucleotide probe (specific competitor);

(d) positive control extract with a 100-fold excess of non-labelled, non-specific, oligonucleotide probe (non-specific competitor).

3   Add 2 μl of 5 × DNA binding buffer to each tube and leave to equilibrate at room temperature for 5 min.

4   Add any unlabelled competing probe, if required, at this point and then add 1 μl of the radiolabelled oligonucleotide consensus probe and agitate the samples at room temperature for 20 min.

5   Add 1 μl of 10 × gel loading buffer to each sample and mix gently using the pipette tip.

6   Load the samples on to the 5% (w/v) non-denaturing polyacrylamide gel and electrophorese at 200 V for approximately 2 h or until the dye has progressed along two-thirds of the length of the gel.

7   Fix the gel for 30 min in gel fixing buffer with one change of buffer after 15 min.

8   Dry the fixed gels on to Whatman 3MM filter paper and monitor DNA binding activity by autoradiography using Hyperfilm MP (Amersham Life Sciences) at −70 °C or by Phosphorimager analysis. Show the free probe in all autoradiographs to demonstrate that it is in excess.

Notes and variations

It is often possible to observe more than one discrete band with retarded mobility caused by the protein–DNA binding event. Whilst these are often non-specific bands, which can be titrated out by increasing the poly dI–dC content of the binding buffer, they can also correspond to DNA–protein interactions of transcription factors of different molecular composition. For example, the NFκB transcription factor dimer can exist as multiple combinations of its family members, which include p50(105), p52(100), p65(RelA), Rel-B, and c-Rel. The composition of such bands can be further investigated by 'super-shift assays' whereby the protein–DNA binding reaction is incubated for a further 10 min in the presence of an antibody directed against a particular subunit of the transcription factor. The antibody–protein–DNA complex migrates slower still through the gel, thus causing an apparent 'super-shift' in its position.

References

1. Callard, R. E. and Gearing, A. J. H. (1994). *The cytokine facts book*. Academic Press, San Diego, CA.

2. Coligan, J. E., Kruisbeek, A. M., Margulies, D. H., Shevach, E. M., and Strober, W. (eds). (1991). *Current Protocols in immunology*. Wiley, New York.

3. Sambrook, J., Fritsch, E. F., and Maniatis, T. (ed.). (1989). *Molecular cloning: a laboratory manual* (2nd edn). Cold Spring Harbor Laboratory Press, NY.

4. Foxwell, B. M. J., Beadling, C., Guschin, D., Kerr, I., and Cantrell, D. (1995). *Eur. J. Immunol.*, **25,** 3041–6.
5. Nisitani, S., Kato, R. M., Rawlings, D. J., Witte, O. N., and Wahl, M. I. (1999). *Proc. Natl. Acad. Sci. USA*, **96,** 2221–6.
6. Page, T. H., Lali, F. V., and Foxwell, B. M. (1995). *Eur. J. Immunol.*, **25**, 2956–60.
7. Sieh, M., Bolen, J. B., and Weiss, A. (1993). *EMBO J.*, **12**, 315–21.
8. Crawley, J. B., Williams, L. M., Mander, T., Brennan, F. M., and Foxwell, B. M. J. (1996). *J. Biol. Chem.*, **271**, 16357–16362.

Chapter 7

Cytokine receptor signalling pathways: PC-specific phospholipase C and sphingomyelinases

Stefan Schütze

Institute of Immunology, University of Kiel, Department of Immunology, Brunswiker Str. 4, D-24105, Kiel, Germany

Martin Krönke

Institute of Medical Microbiology and Hygiene, University of Cologne D-50935 Köln, Germany

1 Introduction

Cytokines constitute a category of regulatory polypeptides that include lymphokines, interleukins, and interferons. They play vital roles as mediators of intercellular communication within the immune system. The two hallmarks of cytokine function are the marked pleiotrophy and redundancy of biological activities. That is, one cytokine may exert many different activities on different cell types, and many of the cytokines display overlapping biological activities. The redundancy of cytokine activities may be based on the utilization of shared key components of intracellular signalling pathways. Receptor-mediated signalling pathways are generally initiated by the binding of intracellular proteins to the cytosolic domains of the receptors. These receptor-associated proteins either function as adaptor proteins for the recruitment of other signalling proteins or are themselves enzymes, e.g. kinases, phosphatases, proteases, or lipases. Receptor-mediated hydrolysis of cellular phospholipids by phospholipases of distinct specificities (phospholipase A_2, phospholipase C, phospholipase D, and sphingomyelinases) is a ubiquitous event of central importance in cellular signalling. Activation of these enzymes results in the generation of early second-messenger molecules which further transduce receptor signals to cytoplasmic target proteins. Since multiple intracellular targets exist for a given second messenger, the cytokine signal may branch off to more complex and diversified intracellular signals.

In this chapter, we focus on the phospholipase C and sphingomyelinase system, generating two distinct early lipid second-messenger molecules, 1,2-diacylglycerol (DAG) and ceramide, respectively.

2 Signalling through phosphatidylcholine-specific phospholipase C

In general, several mechanisms may be responsible for the agonist-induced formation of DAG (see *Figure 1*). One class of phospholipases C act on phosphatidyl-inositol-4,5-bisphosphate (PIP$_2$) to generate DAG and inositol-1,4,5-trisphosphate (IP$_3$). Another distinct phospholipase C species acts exclusively on phosphatidyl-choline (PC), the most abundant phospholipid species of mammalian membranes, which results in the production of DAG and phosphorylcholine (Pchol). A further pathway for the formation of DAG from PC involves phospholipase D (PLD) generating phosphatidic acid (PA), which can be subsequently converted to DAG by PA dephosphorylation (see refs 1 and 2 for a review). Since the second-messenger molecule DAG generated by these different pathways is apparently linked to different biological functions, it is important to discriminate between the phospholipases involved.

A growing number of cytokines and growth factors have been found to elicit intracellular responses by activating phosphatidylcholine-specific phospholipase C (PC-PLC) after binding to their cell-surface receptors (see *Table 1*, and ref. 1 for a review). Among them are IL-1, TNF, IL-3, IL-6, CSF-1, PDGF, EGF, TGF-β, HGF, EPO, endothelin–1, interferons, Fas, CD5, CD14, and LPS.

The functional significance of PC-PLC can be explained by the generation of the lipid second-messenger diacylglycerol, a potent activator of protein kinase C (3) and acidic sphingomyelinase (4). Although PC-PLC activation by receptor triggering is widespread, the physiological role of this phospholipase for cell growth and differentiation is not well understood. In most cell systems, DAG is associated with mitogenesis and proliferation rather than with growth arrest.

DAG is able to mediate transcription-factor NFκB activation (4), pointing to a broad spectrum of biological effects associated with DAG, e.g. regulation of the expression of inflammatory proteins such as IL-1, IL-8, the IL-2 receptor, inducible nitric oxide synthase (iNOS), and the vascular cell-adhesion molecules VCAM-1 and ICAM-1 via NFκB activation (see ref. 5 for a review).

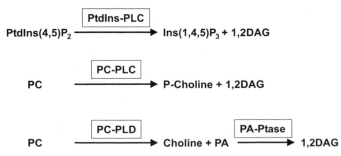

Figure 1 Pathways of diacylglycerol generation.

Table 1 Stimuli signalling through PC-PLC

Interleukin-1 (IL-1)	Hepatocyte growth factor (HGF)
Tumour necrosis factor (TNF)	Erythropoietin (Epo)
Interleukin-3 (IL-3)	Endothelin–1
Interleukin-6 (IL-6)	Interferons α, γ
Colony-stimulating factor (CSF-1)	Fas (CD95, APO-1)
Platelet-derived growth factor (PDGF)	CD5
Epidermal growth factor (EGF)	CD14
Transforming growth factor (TGF-β)	Lipopolysaccharide (LPS)

The evaluation of the functional significance of PC-PLC-generated DAG is unfortunately hampered by the fact that PC-PLC has still not been purified or cloned and that PC-PLC-specific antibodies are also not available. However, evidence for the important role of PC-PLC was provided by a number of reports employing a PC-PLC specific inhibitor, the tricyclodecan-9yl-xanthogenate D609. D609 efficiently prevents the production of DAG from PC in the low micromolar range (4, 6, 7), leaving other enzymes like PLA$_2$, PI-PLC, PC-PLD, PI3-kinase, PKC, tyrosine kinases, phosphatases, and neutral sphingomyelinase unaffected (4, 6, 7). Although it was reported that D609 also blocks PC-PLD activity (8), other investigations have confirmed the PC-PLC specificity. The observed inhibitory effects of D609 on PLD may be a secondary effect of the inhibition of PC-PLC, since it is well known that PLD is regulated by DAG-sensitive protein kinase C (PKC) (2) and that PC-PLC and PLD enzymatic activities are coupled. In the case of D609 inhibition of PLD, the blockade in PC-PLC-generated DAG will result in a lack of PKC activation and subsequently the lack of PLD stimulation (ref. 9, and our own unpublished observations).

As to a functional role of PC-PLC, in a recent paper by Machleidt *et al.* 1996 (10), we showed that D609 blocks the cytotoxic action of tumour necrosis factor (TNF) on L929 and Wehi164 cells.

In vivo, D609 prevented both the expression of adhesion molecules in the pulmonary vasculature and the accompanying leucocyte infiltration in TNF-treated mice. Most strikingly, D609 protected BALB/c mice from lethal shock induced either by TNF, lipopolysaccharide (LPS), or staphylococcal enterotoxin B (SEB). D609 did not inhibit the SEB-induced release of IL-1, TNF, or IFN-γ in SEB-treated animals. The findings of Machleidt *et al.* were later confirmed in an independent study by Tschaikowsky *et al.* (11), in which D609-treated mice showed an enhanced survival from endotoxin shock. In this study it was also demonstrated that D609 reduced the release of IL-1β, IL-6, and nitric oxide *in vivo*, leaving TNF serum levels unaltered. It is important to note that PC-PLC is also involved in the signalling pathways of other inflammatory cytokines like IL-1 and IFN-γ and LPS itself (see *Table 1*), which synergize with TNF to produce a lethal shock syndrome. Thus D609 will concomitantly antagonize the pathological action of all these mediators in the progression of a shock syndrome. The protective mechanism of D609 is different from other known inhibitors of

LPS/SEB-mediated lethal shock, such as chlorpromazine, adenosine kinase inhibitors, or tyrphostin all of which diminish cytokine production. Together, these findings imply that PC-PLC is an important mediator of the pathogenic action of TNF and other inflammatory cytokines—suggesting that PC-PLC might serve as a novel target for anti-inflammatory cytokine antagonists, especially as it can block the action of already secreted proinflammatory cytokines.

Intriguingly, DAG-regulated protein kinase C is involved in antiapoptotic signalling and, since one of the major biological functions of ceramide is apoptotic signalling, DAG may be thought of as a lipid mediator that contraregulates the function of ceramide.

3 Signalling through sphingomyelinases

In addition to the lipid second-messenger molecule DAG, the sphingolipid-derived second messenger ceramide has been implicated in various important pathways of a growing number of diverse stimuli, summarized in *Table 2*.

Ceramide is a common intracellular second messenger for various cytokines, growth factors, Fas/APO-1, and other stimuli such as chemotherapeutic drugs and stress factors (see refs 12–17 for reviews). The biological responses to ceramide range from the induction of proliferation and differentiation to cell-cycle arrest. The most prominent, yet controversial, role of ceramide is the induction of apoptosis in various cell types (reviewed in refs 12–14, 17).

The production of ceramide is mediated either by *de novo* synthesis involving ceramide-synthase located in the endoplasmic reticulum, or by hydrolysis of sphingomyelin engaging sphingomyelinase (SMase) (see *Figure 2*).

Ligand binding to the p55 TNF receptor (TR55), interleukin-1 receptor 1 (IL-1 RI) and the Fas receptor results in the activation of two SMases: a plasma membrane-bound neutral SMase (N-SMase) as well as an endolysosomal acid SMase (A-SMase) (4, 18–20). Each type of SMase generates the second messenger ceramide, albeit with different kinetics and, most importantly, at different intracellular locations. Structure–function analysis of the cytoplasmic domain of TR55, revealed that specific TNF receptor domains link to the respective SMases and to diverse signal-

Table 2 Stimuli signalling through ceramide

Tumour necrosis factor (TNF)	Complement C5b
Interleukin-1β	Daunorubicin
Fas (CD95, APO-1)-ligand	Etoposide
CD28	Vincristine
Interferon-γ	Corticosteroids
Nerve growth factor (NGF)	REAPER
Platelet-derived growth factor (PDGF)	Oxidative stress
Angiotensin II	Serum starvation
Progesterone	Ultraviolet light
Vitamin D_3	Ionizing radiation

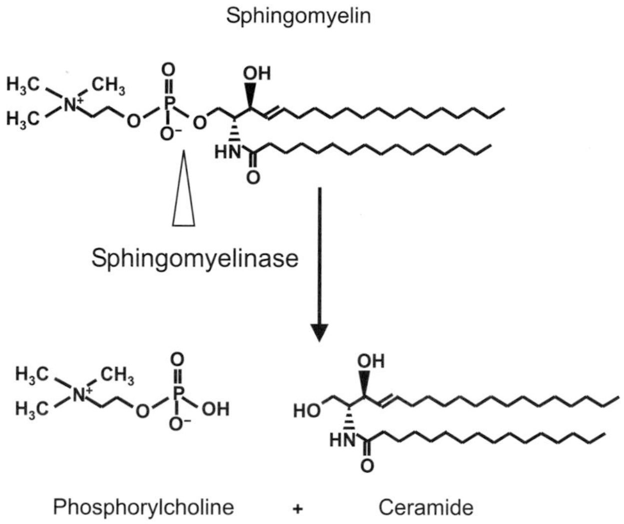

Figure 2 Production of ceramide from sphingomyelin cleavage.

ling pathways (18). N-SMase activation is coupled to a neutral sphingomyelinase activation domain (NSD), via the adaptor protein FAN (21). The domain of TR55 activating A-SMase corresponds to the death domain which binds the adaptor protein TRADD (TNF receptor-associated protein with death domain), that in turn recruits a further protein, FADD (Fas-associated protein with death domain), to activate caspase 8 and A-SMase (22). Fibroblasts derived from FADD-deficient mice are defective in TNF-induced A-SMase activation (23). The apoptotic signalling cascade downstream of ceramide may involve caspase 3 (CPP32) (24). A role for A-SMase in transmitting apoptotic signals in response to Fas/CD95, γ-irradiation, TNF, and LPS has been suggested by several reports (20, 25–29). However, direct evidence linking ceramide signalling to specific effector elements of the apoptotic response has yet to be provided. The identification of direct target proteins of ceramide will help our understanding of the role of ceramide during apoptosis and other cellular responses. The scheme depicted in *Figure 3* summarizes the current state of knowledge of those enzymes found to be responsive to ceramide, and which are possibly involved in TNF signalling. These include a ceramide-activated protein kinase (CAPK) (30), an enzyme suggested to be related to or identical with the kinase suppressor of Ras (KSR) (31). Further ceramide-responsive proteins include protein phosphatases 2A (CAPP) (32, 33) and protein phosphatase A1 (34), protein kinase C ξ (35), protein kinase Raf-1 (36, 37), the stress-activated/c-jun N-terminal protein kinase (JNK) (38–40), and a CPP32 (caspase 3)-like apoptotic protease (41–44). The functional link of one of these enzymes to one of the two SMases is not yet fully resolved, but is the subject of intensive investigations.

By employing ceramide-affinity chromatography and D-erythroceramide-based, photo-crosslinking we have recently identified cathepsin D (CTSD) as a novel ceramide-binding protein (44, 45). Ceramide binding enhances the

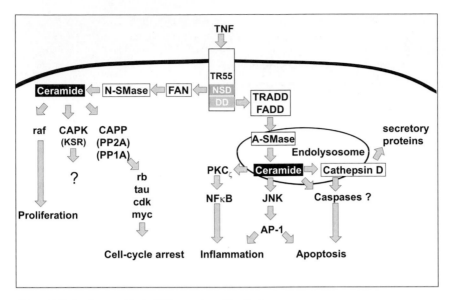

Figure 3 Role of ceramide in TNF receptor signalling.

enzymatic activity of CTSD and induces the maturation of the prepro-CTSD isoform to generate the active CTSD isoforms. CTSD is endolysosomally active and thus co-localizes with the A-SMase. Acidic SMase-deficient cells derived from patients with Niemann–Pick disease show decreased CTSD activity; however, activity was restored by transfection with A-SMase cDNA. Ceramide accumulation in cells derived from A-ceramidase defective Farber patients correlates with enhanced CTSD activity. These findings suggest that A-SMase-derived ceramide targets endolysosomal CTSD.

The protease is involved in the proteolytic activation of regulatory proteins and, most interestingly, has been recently implicated in mediating apoptosis in response to TNF, IFN-γ, CD95 (46), chemotherapeutic agents like etoposide and adriamycin (47), serum deprivation (48, 49), and oxidative stress (50, 51). Thus CTSD may link A-SMase to the secretory pathway and to apoptotic signalling events.

In this chapter methods are described for the detection of agonist-induced activation of PC-PLC and SMases and for the production of the lipid second messengers 1,2-DAG and ceramide.

4 Phosphatidylcholine-specific phospholipase C

In general, cytokine activation of PC-PLC in whole cells can be monitored by estimating increases in the total amounts of the cleavage products, 1,2-diacylglycerol (DAG) and phosphorylcholine (Pchol), and by estimating the concomitant substrate degradation, i.e. the reduction of phosphatidylcholine (PC). However, several problems have to be addressed. First, DAG production can also result from the cleavage of other phospholipid sources (PIP_2, PE) or from the dephosphorylation of phosphatidic acid (PA). Second, the other cleavage product, Pchol, can also be produced by the sphingomyelinase (SMase)-mediated cleavage of sphingomyelin (SM) or by the phosphorylation of choline by choline kinase.

4.1 Detection of 1,2-DAG production from PC in metabolically labelled cells

Several methods have been successfully used to document changes in DAG levels in haemopoietic cells: for instance, by total mass analysis of dried organic extracts performed either by capillary gas chromatography (52), or high-performance liquid chromatography (HPLC) (53). DAG mass levels can also be estimated by the phosphorylation of neutral lipid extracts employing DAG-kinase from *Escherichia coli* and [γ-^{32}P]ATP incorporation (54). This assay system is commercially available (Amersham). A second method is based on metabolically labelling the cellular PC pools with the precursors [^{3}H]glycerol, [^{3}H]myristic acid (55), 1-[^{3}H]alkyl-2-lyso-glycero-3-phosphatidylcholine (56), or 1-[^{14}C]palmitoyl-2-lyso-3-phosphatidylcholine (57). Radiolabelled DAG is subsequently analysed after extraction of neutral lipids by high-performance, thin-layer chromatography (HPTLC). This method allows a considerable number of samples to be processed; in addition, the total amount of material applied on TLC can be controlled (cholesterol or triglycerols as internal standards for neutral lipids extracted). In the following section, DAG analysis will be described in cells labelled with 1-[^{14}C]palmitoyl-2-lyso-3-phosphatidylcholine (as described in ref. 57), a rather selective labelling of PC based on the view that agonist-induced PC-cleavage is believed to occur in specific, hormone-sensitive pools (58).

Protocol 1

Metabolic cell labelling for the detection of 1,2-DAG and cell stimulation

Equipment and reagents

- Hanks' balanced salt solution (HBSS)
- Serum-free growth medium
- Bovine serum albumin (BSA)
- Dry ice
- 1-[^{14}C]palmitoyl-2-lyso-3-phosphatidylcholine (Sp. Act. 56.8 mCi/mmol, Amersham)
- Methanol

Method

Cells should be serum-starved for 4 h prior to stimulation with agonists.

1 Wash cultured cells at least three times in HBSS and incubate in serum-free growth medium, supplemented with 2% BSA for 2 h at 37 °C.

2 Add 1 µCi/ml final concentration of 1-[^{14}C]palmitoyl-2-lyso-3-phosphatidylcholine and incubate for a further 2 h in serum-free medium at 37 °C.

Protocol 1 continued

3   Wash the cells with HBSS and resuspend to 10^7 cells/ml for subsequent agonist stimulation.

4   Treat 0.5-ml aliquots of the cell suspensions with the agonist of choice for various times at 37 °C. Stop treatment by immersing the sample tubes in methanol/dry ice for 10 sec followed by centrifugation for 20 sec at 4 °C in a microcentrifuge.

5   Discard the supernatants into a radioactive waste container and resuspend the cell pellets in 1 ml of cold methanol.

Protocol 2 describes two procedures that either allow for the separation of neutral lipids, phospholipids, and water-soluble components according to the method of Bligh and Dyer (59) (Part A), or a more rapid procedure, which separates an organic phase containing both neutral and phospholipids from the aqueous phase (Part B).

Protocol 2

Lipid extraction

Equipment and reagents

- Methanol
- Chloroform
- 10 ml glass tubes
- Waterbath sonicator
- Vortex mixer
- Nitrogen gas
- Hexane

A. Separation of neutral and phospholipids and water-soluble components

1   Transfer samples that have been resuspended in 1 ml methanol (*Protocol 1*) to 10 ml glass tubes containing 1 ml H_2O and 1.5 ml methanol. Sonicate the tubes for 5 min in a waterbath sonicator.

2   Add 1.25 ml of chloroform, briefly vortex, and centrifuge for 10 min at 6000 *g*.

3   Save the supernatants in new glass tubes. Resuspend the pellets in 1 ml of H_2O, 2.5 ml of methanol, and 1.25 ml of chloroform. Repeat the sonication for 5 min and centrifuge for 5 min at 4000 *g* at 4 °C.

4   Combine the two supernatants, add 2.5 ml H_2O and 2.5 ml chloroform, vortex, and centrifuge for phase-separation for 5 min at 4000 *g* at 4 °C. The pellets are discarded.

5   Transfer and save the lower organic phase in new glass vials. Add 4 ml chloroform to the residual aqueous phase, mix, centrifuge for phase-separation as above, and combine the lower, organic phase with the first.

6   Dry-down the chloroform phases under nitrogen.

7   Resuspend the dried samples, for separation of neutral lipids and phospholipids, in 4 ml of hexane/3 ml methanol, mix, centrifuge as above for phase-separation. The methanol phase contains phospholipids, the hexane phase contains neutral lipids.

Protocol 2 continued

8 Transfer and save the lower methanol phase into new glass vials, add 3 ml methanol and 2 ml hexane to the first residual hexane phase, mix, and centrifuge for phase-separation (step 4).

9 Transfer the lower methanol phase to the first from step 8, repeat the methanol washing of the hexane phase two more times.

10 Dry-down the separated methanol and hexane fractions under nitrogen.

11 Resuspend the phospholipids from the methanol fraction and neutral lipids from the hexane fraction in 50 μl of chloroform/methanol (9:1) for TLC analysis.

B. Separation of total lipids from the aqueous phase

1 For the rapid isolation of total lipids, follow Part A, steps 1–6.[a]

2 Take the dried chloroform phase containing both neutral and phospholipids for TLC separation analysis.

[a] The separation of neutral and phospholipids by hexane/methanol extraction prior to TLC analysis of DAG or phospholipids, respectively, can be omitted, because the solvent systems adopted for TLC analysis of DAG and PL will also separate the two lipid fractions.

Protocol 3

Thin-layer chromatography for the detection of 1,2-DAG

Equipment and reagents

- HPTLC silica gel 60 plates (Merck) and TLC chamber
- Methanol
- Benzene/ethylacetate (6.5:3.5)
- Neutral lipid standards: 1,2-DAG, 1,3-DAG, MAG, TG, cholesterol
- Chloroform
- Kodak XAR-films *or* Laser densitometer *or* Liquid scintillation counter
- Phosphorimager equipment
- Iodine vapour, charring densitometry apparatus and reagents

Method

1 Pre-run the plates in a solvent system composed of methanol:chloroform (1:1).

2 Dry the plates at 80 °C for 30 min.

3 Pre-equilibrate the TLC chambers for 1 h at room temperature with the solvent system to be used for lipid separation. To detect 1,2-DAG, separate neutral lipids using a solvent system of benzene/ethylacetate (6.5:3.5).

4 Resuspend the dried samples in 50 μl chloroform/methanol (9:1). Spot the samples on the HPTLC plate in strips of 5–8 mm for each sample about 2 cm from the base of the plate and allow to dry. Load samples containing equal amounts of radio-activity. Apply 1,2-DAG, 1,3-DAG, monoacylglycerol (MAG), triglycerol (TG), and cholesterol as standards.

5 Place the TLC plate into the chamber and allow the plates to develop at room temperature until the front is within 1 cm from the top.

Protocol 3 continued

6   Dry the plates in an oven or by using a fan, and expose to Kodak XAR-films at room temperature for 1–2 weeks or develop the plates in a phosphor-imaging system after 2–4 days.

7   Analyse the autoradiographs and quantitate DAG by two-dimensional scanning. Alternatively, scrape the corresponding spots from the TLC plates and determine the radioactivity in a scintillation counter. (A typical distribution of radiolabelled DAG visualized in a FUJI-X 1000 Bioimager (Raytest) is shown in *Figure 4A*.) Visualize the lipid standards by exposure to iodine vapour, Coomassie-Blue staining (Section 5.1.4), or by charring densitometry (*Protocol 5*).

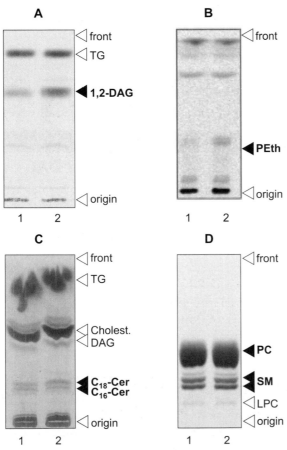

Figure 4 TLC—resolution of lipids. U937 cells (in A, C, D) or mast cells (in B) were left untreated as controls (lanes 1) or stimulated (lanes 2) with bacterial PC-PLC (A), in the presence of 1% ethanol with NGF (B), or with TNF (C, D). Lipids were extracted according to *Protocol 2* and analyzed by TLC using the indicated solvent systems: **A)** ^{14}C-labeled DAG, solvent: benzene/ethylacetate (6.5/3.5), autoradiography. **B)** ^{14}C-labeled PEth, solvent: the organic phase of ethylacetate/acetic acid/H_2O (110:20:110), autoradiography. **C)** Ceramide, solvent: system: dichloromethane/methanol/acetic acid (100:2:5) charring densitometry. **D)** ^{14}C-labeled phospholipids, solvent: chloroform/methanol/acetic acid/H_2O (100:60:20:5), autoradiography.

142

4.2 Analysis of phosphatidylcholine (PC)

In contrast to labelling at equilibrium with [^{14}C]choline for 48 h, the rather selective labelling of agonist-sensitive PC pools with lyso-PC for 2 h will more readily allow the detection of cytokine-induced changes in PC.

4.2.1 Metabolic labelling, cell stimulation, and extraction of PC

Agonist-induced changes in membrane PC content can be monitored after short-term labelling with lyso-PC as described for the detection of 1,2-DAG (see Section 4.1).

(a) Follow *Protocol 1* for radioactive labelling and cell stimulation. Instead of using 1-[^{14}C]palmitoyl-2-lyso-3-phosphatidylcholine, similar results can be obtained with L-lyso-3-phosphatidyl [methyl-^{14}C]choline (1 mCi/ml, Sp. Act. 22 mCi/mmol).

(b) Extract phospholipids according to *Protocol 1*. The dried methanol phase is used for PC analysis by TLC. Alternatively, follow *Protocol 2* and use the whole organic phase for TLC analysis.

4.2.2 Thin-layer chromatography for the detection of PC

After preparing the silica-60 TLC plates and equilibrating the chambers with the solvent system containing ($CHCl_3/CH_3OH/CH_3COOH/H_2O$) (100:60:20:5), the dried samples are resolved in chloroform/methanol (9:1) and spotted on the TLC plates. Phospholipids containing PC are separated in the solvent system above. Lyso-PC, PC, phycoerythrin (PE), PI, phosphatidic acid (PA), and SM are used as standards and visualized by iodine vapour, Coomassie-Blue staining, or charring densitometry (see *Protocol 5*). Radiolabelled phospholipids are examined by auto-radiography followed by 2D laser scanning. Spots can also be scraped off the plates followed by liquid scintillation counting. A typical distribution of phospholipids labelled with [methyl-^{14}C]choline is presented in *Figure 3D*.

4.3 Detection of phosphorylcholine (Pchol)

The second cleavage product of PC-PLC, phosphorylcholine, can be detected after radiolabelling cells with either [methyl-^{14}C]choline or lyso-phosphatidyl-[methyl-^{14}C]choline. Cells should be kept serum-starved for 4 h prior to stimulation with agonists.

Protocol 4

Thin-layer chromatography for the detection of phosphorylcholine (Pchol)

Equipment and reagents

- Equipment and reagents as *Protocol 1*
- Methanol
- Solvent system: $CH_3OH/0.5\%$ NaCl/NH_4OH (100:100:2)

- TLC plates and chambers
- Autoradiography systems
- Standards: Phosphorylcholine, choline, acetylcholine, glycerophosphoryl-choline

Protocol 4 continued

Method

NB: Choline metabolites are prepared from the aqueous phase obtained from *Protocol 1*.

1   Follow steps 1–4, *Protocol 1*

2   Transfer the upper, aqueous phase to a new vial, and freeze-dry the samples.

3   Dissolve the freeze-dried samples in 50 μl methanol.

4   Prepare the TLC plates and TLC chambers as described in *Protocol 3*.

5   Separate the water-soluble components containing Pchol on silica-60 TLC plates using the solvent system CH_3/0.5% $NaCl$/NH_4OH (100:100:2). Use phosphorylcholine, glycerophosphorylcholine, acetylcholine, and choline as standards.

6   Evaluate the radiolabelled choline metabolites by autoradiography and visualize the standards as described in *Protocol 3*.

To detect water-soluble choline metabolites, cells are labelled for 48 h with [methyl-^{14}C]choline (1 μCi/ml, Sp. Act. 56.4 mCi/mmol), which will result in at-equilibrium labelling. Alternatively, cells can be labelled for 2 h with L-lyso-3-phosphatidyl-[methyl-^{14}C]choline (Sp. Act. 22 mCi/ml).

4.4 Estimation of phosphatidylcholine-specific phospholipase D (PC-PLD) activity

PC-PLD activity can be monitored directly by estimating changes in its cleavage product phosphatidic acid (PA), or by determining the formation of labelled phosphatidylethanol (PEth), which is expected from PLD's transphosphatidyl-ation activity in the presence of ethanol (61). A possible involvement in DAG formation can be tested by using a PA-phosphohydrolase inhibitor, propranolol (62).

4.4.1 Estimation of phosphatidic acid

(a) For detecting PA, label cells with 1-[^{14}C]palmitoyl-2-lyso-3-phosphatidylcho-line (1 μCi/ml (Sp. Act. 56.8 mCi/mmol) for 2 h as described in *Protocol 1* for the detection of DAG.

(b) Extract phospholipids according to *Protocol 2* using the dried methanol phase (Part A) or the whole organic phase (Part B) for PA-analysis by TLC, respect-ively.

After preparing the silica-60 TLC plates using acetone as the solvent, the TLC chambers are equilibrated with the solvent according to *Protocol 3*. The dried samples are resolved in chloroform/methanol (9:1), spotted on the TLC plates, and phospholipids containing PA are separated in a solvent system containing ($CHCl_3$/CH_3OH/CH_3COOH) (65:15:5). PA, PC, PI, PE, SM, and lyso-PC are used as standards and visualized by iodine vapour, Coomassie-Blue staining, or charring densitometry (see *Protocol 5*). Radiolabelled phospholipids are examined by auto-radiography followed by 2D laser scanning, phosphoimaging, or liquid scintil-lation counting of scraped-off spots.

4.4.2 Transphosphatidylation in the presence of ethanol

Cells are labelled with 1-$[^{14}C]$palmitoyl-2-lyso-3-phosphatidylcholine (1 μCi/ml (Sp. Act. 56.8 mCi/mmol, Amersham) for 2 h as described for the detection of PA (Section 4.4.1.). Prior to agonist stimulation, cells are pre-incubated for 2 min with 1% ethanol. Phospholipid extraction is performed as described for PA (Section 4.4.1.). For separating phosphoethanol (PEth) from PA and other phospholipids, TLC is performed in a solvent system consisting of the organic phase of ethylacetate/acetic acid/H_2O (110:20:110). A typical resolution of $[^{14}C]$PEth by autoradiography is depicted in *Figure 4B*.

4.5 Inhibition of PC-PLC by D609

To evaluate the biological functions of PC-PLC in cytokine signalling, a rather selective PC-PLC inhibitor, the xanthogenate D609, has been employed in many studies. This compound does not block PI-PLC, PC-PLD, PLA$_2$, SMase, PKC, Tyr-PK, PI-kinase (4, 6, 7, 11). The effective dosage may depend on the cell type. Cells were usually treated with 20–100 μg/ml D609 for 30 min prior to agonist stimulation. Since non-specific side reactions cannot be ruled out completely, in each individual experiment the effect of D609 on DAG production as well as on other, not PC-PLC-related, systems responsive to the individual stimuli has to be checked.

5 Sphingomyelinases

The production of ceramide is mediated either by *de novo* synthesis involving ceramide synthase located in the endoplasmic reticulum, or by the hydrolysis of sphingomyelin (SM) engaging sphingomyelinases (54). Activation of the SM-hydrolysis pathway in whole cells can be monitored either by measuring the reduction in the SM content in radiolabelled cells, or the production of the cleavage products ceramide and phosphorylcholine (Pchol). Ceramide can be easily detected from labelled cells by TLC and autoradiography of radioactive ceramide, or from unlabelled cells by mass analysis using DAG kinase from *E. coli* (54). A more advanced and sensitive methodology is the use of electrospray mass spectrometry, which will separate distinct molecular ceramide subspecies (64, 65). This technique, however, requires the appropriate equipment. Pchol is measured as described for the analysis of PC-PLC action (Section 4.3). Pchol can be produced by various different pathways *in vivo*, involving either PC-PLC, SMase, or choline kinase. Therefore, we use this cleavage product as a read-out for SMase-activity only in a micellar assay system with defined, exogenous substrates (see Section 5.2.).

5.1 Detection of SM hydrolysis in metabolically labelled cells

Hydrolysis of SM in whole cells can be measured either after radiolabelling the SM pools, using [methyl-^{14}C]choline for labelling the choline moiety of the sphingolipid as described below, or by analysing total SM mass by charring densitometry (*Protocol 5*).

5.1.1 Cell culture and metabolic labelling for the analysis of SM

For the detection of SM, cells are metabolically labelled for 48 h with [methyl-^{14}C]choline (1 μCi/ml, Sp. Act. 56.4 mCi/mol) in culture medium containing 5% FCS. Prior to agonist stimulation, cells are washed serum-free with HBSS and kept serum-starved for 4 h in medium supplemented with 2% BSA as described in Section 4.1.

Cell stimulation is performed as described in *Protocol 1*, steps 4 and 5.

5.1.2 SM extraction

Cell pellets (*Protocol 1*, step 5) are resuspended in cold methanol and SM is extracted with the methanol (phospholipid) fraction described in *Protocol 2*, Part A. Alternatively, SM can also be resolved from total lipids from the chloroform fraction (Part B).

5.1.3 Thin-layer chromatography for the detection of SM

Dried lipids are resolved in chloroform/methanol (9:1) and spotted on to silica-60 TLC plates, which have been pre-treated with chloroform/methanol as described in Section 4.1. To separate SM from other phospholipids, the solvent system ($CHCl_3/CH_3OH/CH_3COOH/H_2O$) (100:60:20:5) is employed. SM, PC, PA, PI, PE, and lyso-PC are used as standards and visualized by iodine vapour, Coomassie-Blue staining, or charring densitometry (see *Protocol 5.*). Radiolabelled phospholipids are examined by autoradiography followed by 2D laser scanning, using a phosphor-imaging system, or by scraping off the spots and liquid scintillation counting. A typical resolution of [^{14}C]SM is shown in *Figure 3D*.

After TLC separation, phospholipids can be visualized and quantitated by charring the plates with a cupric reagent (60) (*Protocol 5*).

Protocol 5

Charring densitometry

Equipment and reagents

- 2D scanner densitometry system (Molecular Dynamics Personal Densitometer)
- Glass chamber
- 10% copper sulfate in 8% aqueous phosphoric acid
- Heating plate (180 °C)

Method

1   After separating the samples by TLC, remove the plates from the TLC chamber and dry the plates for 10 min at 180 °C.

2   Let the plates cool down to room temperature and expose for 15 sec to a solution of 10% copper sulfate in 8% aqueous phosphoric acid in a vertical glass chamber.

3   Transfer the plates back to the heater and dry for 2 min. at 110 °C.

4   Char at 175 °C for approximately 5–10 min until brown bands become visible. Remove the plates from the heater before the background becomes dark.

5   Non-radioactive lipids can be analysed by scanning the charred TLC plates by 2D laser densitometry. For quantitating radiolabelled lipids, scrape the charred bands off the plates and determine the radioactivity by liquid scintillation counting.

5.1.4 Coomassie Brilliant Blue staining

After separation by TLC, lipids can be visualized by staining the plates in a solution of 35% methanol, 100 mM NaCl, and 0.03% Coomassie Brilliant Blue R250 (63). Plates are destained in 35% methanol, 100 mM NaCl. Charring densitometry is recommended for the quantitative analysis of the lipids (*Protocol 5*).

5.2 Detection of ceramide

The neutral lipid cleavage product of SMases, ceramide, can either be measured by total mass analysis, employing charring densitometry of TLC plates as described in Section 5.1.3, or by total mass determinations using DAG-kinase (54; for a discussion of this method see Kolesnick and Hannun (17), Perry Hannun (66), and Watts *et al.* (67). For this approach, the DAG-kinase assay system from Amersham may be used following the procedure described in ref. 54. The most sensitive and informative way to analyse even molecular ceramide subspecies is to employ negative electrospray-ionization mass spectroscopy (64, 65), which, however, requires the appropriate equipment. An easier way of identifying changes in ceramide levels, based on the determination of radiolabelled ceramide isolated from metabolically labelled cells and analysis by TLC, is described below.

5.2.1 Metabolic labelling

To detect ceramide, the SM pool of serum-starved cells are metabolically labelled with 1-[^{14}C]palmitoyl-2-lyso-3-phosphatidylcholine (1 μCi/ml, Sp. Act. 56.8 mCi/mmol) for 2 h as described for the detection of DAG (Section 4.1). Cell stimulation is performed as described in *Protocol 1*.

5.2.2 Ceramide extraction and TLC analysis

Cell pellets are resuspended in cold methanol and neutral lipids extracted. Ceramides are recovered in the hexane fraction following the lipid extraction

Table 3 Solvent systems for lipid analysis on TLC

Lipid metabolite	Solvent system	Ratio
1,2-Diacylglycerol	Benzene:ethylacetate	65:35
Phosphatidylethanol	Organic phase of: ethylacetate:chloroform:H_2O	110:20:110
Ceramide	Dichloromethane:methanol:acetic acid	100:2:5
Phosphorylcholine	Methanol:0.5% NaCl:NH_4OH	100:100:2
Phosphatidylcholine Phosphatidic acid Sphingomyelin	Chloroform:methanol:acetic acid:H_2O	100:60:20:5

described in *Protocol 2*, Part A. Alternatively, ceramides can also be recovered from the total lipid (chloroform) fraction when Part B in *Protocol 2* is followed.

To separate ceramide from other neutral lipids, dried samples are resolved in chloroform/methanol (9:1) and spotted on to silica-60 TLC plates, which have been pre-treated as described in Section 4.1.4. The solvent system (dichloromethane/methanol/acetic acid (100:2:5) is used for the detection of ceramide. C-16 ceramide, 1,2-DAG, MAG, and TG are employed as standards and visualized by iodine vapour, Coomassie staining, or charring densitometry (see Section 5.1, *Protocol 5*). Radiolabelled ceramide is examined by autoradiography followed by 2D laser scanning or using a phosphor-imager system. A typical resolution of ceramide by charring densitometry is shown in *Figure 4C*.

The solvent systems used for separating the various lipid metabolites is summarized in *Table 3*.

5.3 Estimation of SMase activity *in vitro*

Agonist-stimulated SMase activities can also be estimated by *in vitro* enzymatic micellar assays of detergent-solubilized SMase preparations (18). A selective estimation of neutral and acidic SMases activities can be estimated with this method. Acidic SMase is a rather stable enzyme and is insensitive to proteolysis. Thus acidic-SMase preparations can even be frozen down and thawed whilst retaining full activity. Neutral SMase, in contrast, is highly susceptible to proteolytic inactivation. Therefore, the preparation of this enzyme has to be performed in the presence of various inhibitors, and enzymatic assays should be performed on the same day.

5.3.1 Cell stimulation

(a) Cells are serum-starved for 4 h prior to agonist stimulation. No radioactive labelling is required.

(b) Aliquots of 0.5 ml cell suspensions (1×10^7 cells/ml) are stimulated with the agonist of choice for various times at 37 °C. Treatments are stopped by immersing the sample tubes in methanol/dry ice for 10 sec, followed by centrifugation at 4 °C in a microcentrifuge. Since N-SMase activation is a rapid and transient event, it is crucially important to keep the samples at 4 °C throughout the following extraction procedure.

Protocol 6

Extraction of neutral SMase

Equipment and reagents

- Buffer A: 20 mM Hepes pH 7.4, 10 mM MgCl$_2$, 2 mM EDTA, 5 mM DTT, 0.1 mM Na$_3$VO, 0.1 mM Na$_2$MoO$_4$, 30 mM p-nitrophenylphosphate, 10 mM

- β-glycerophosphate, 1 mM PMSF, 10 μM leupeptin, 10 mM pepstatin, 750 μM ATP, 0.2% NP-40
- 18-gauge needle and syringe

Method

1 Resuspend the cell pellets in ice-cold buffer A.

2 After 5 min at 5 °C, homogenize the cells by repeatedly squeezing them through an 18-gauge needle.

3 Remove cell debris and nuclei by low speed centrifugation (800 g) for 5 min.

4 Use the supernatants containing NP-40-solubilized enzymes for the *in vitro* assay of agonist-stimulated N-SMase activities.

Protocol 7

Assay for neutral SMase activity

Equipment and reagents

- Reaction buffer: 20 mM Hepes (pH 7.4), 1.0 mM MgCl$_2$, 2.25 μl (1 nmol) [N-methyl-^{14}C]sphingomyelin (0.2 μCi/ml, Sp. Act. 56.6 mmol)
- Chloroform/methanol (2:1 (v/v))

- Vortex mixer
- Microcentrifuge and tubes
- Liquid scintillation counter

Method

NB: To allow linear enzymatic reactions, the amount of [^{14}C]SM hydrolysed should not exceed 10% of the total amount of radioactive SM added to the assay.

1 Add 50 μg of protein to the reaction buffer in a microtube in a total volume of 50 μl.

2 Incubate the samples for 2 h at 37 °C.

3 Stop the reactions by adding 800 μl chloroform/methanol (2:1), vortex-mix, add 250 μl of H$_2$O, vortex-mix, and centrifuge for 2 min at full speed in a microcentrifuge for phase separation.

4 Analyse 200 μl of the aqueous upper phase of each sample for ^{14}C-labelled Pchol by liquid scintillation counting.

Protocol 8

Extraction of acidic SMase

Equipment and reagents

- 0.2% Triton X-100 or 0.05% NP-40
- 18-gauge needle and syringe

Method

1 Resuspend cell pellets in 200 μl of 0.2% Triton X-100. Alternatively, 0.05% NP-40 can be used as detergent.

2 After 5 min at 5 °C, homogenize the cells by repeatedly squeezing them through an 18-gauge needle.

3 Remove cell debris and nuclei by centrifugation (800 g) for 5 min.

4 Use the supernatants, containing Triton X-100-solubilized enzymes for the *in vitro* assay of A-SMase activities.

Protocol 9

Assay for acidic SMase

Equipment and reagents

- Reaction buffer: 250 mM Na-acetate pH 5.0, mM EDTA, 2.25 μl (1 nmol) of [*N*-methyl-^{14}C]sphingomyelin (0.2 μCi/ml, Sp. Act. 56.6 mmol)
- Liquid scintillation counter
- Chloroform/methanol (2:1)
- Vortex mixer
- Microcentrifuge and tubes

Method

1 Add 50 μg of protein to the reaction buffer (total volume of 50 μl) in a microcentrifugation tube.

2 Incubate the samples for 2 h at 37 °C.

3 Stop the reactions by adding 800 μl chloroform/methanol (2:1), vortex-mix and add 250 μl H$_2$O, vortex-mix, and centrifuge for 2 min at full speed in a microcentrifuge for phase separation.

4. Analyse 200 μl of the aqueous upper phase of each sample for ^{14}C-labelled Pchol by liquid scintillation counting.

Acknowledgement

The work of our laboratory was supported by grants from the Deutschen Forschungs-gemeinschaft (SFB 415).

References

1. Schütze, S. and Krönke, M. (1994). In *Signal-activated phospholipases* (ed. M. Liskovitch), pp. 101–24. R. G. Landes, CRC Press, Florida, USA.
2. Exton, J. H. (1994). *Biochim. Biophys. Acta* **1212**, 26–42.
3. Newton, A. C. (1995). *J. Biol. Chem.* **270**, 28495–8.
4. Schütze, S., Potthoff, K., Machleidt, T., Berkovic, D., Wiegmann, K., and Krönke, M. (1992). *Cell* **71**, 765–76.
5. Baeuerle, P.A. and Henkel, T. (1994) *Annu. Rev. Immunol.* **12**, 141–79.
6. Amtmann, E. (1996). *Drugs Exp. Clin. Res.* **XXII**, 287–94.
7. Cai, H., Erhardt, P., Troppmair, J., Diaz-Meco, M. T., Sithanandam, G., Rapp, U. R., Moscat, J., and Cooper, G. M. (1993). *Mol. Cell Biol.* **13**, 7645–51.
8. Kiss, Z. and Tomono, M. (1995). *Biochim. Biophys. Acta* **1259**, 105–8.
9. Han, J. S., Hyun, B. C., Kim, J. H., and Shin, I. (1999). *Arch. Biochem. Biophys.* **367**, 233–9.
10. Machleidt, T., Krämer, B., Adam, D., Neumann, B., Schütze, S., Wiegmann, K., and Krönke, M. (1996) *J. Exp. Med.* 184, 725–33.
11. Tschaikowsky, K., Schmidt, J., and Meisner, M. (1998). *J. Pharmacol., Exp. Ther.* **285**, 800–4.
12. Hannun, Y. A. and Obeid, L. M. (1995). *TIBS* **20**, 73–7.
13. Kolesnick, R. N. and Krönke, M. (1998). *Annu. Rev. Physiol.* **60**, 643–65.
14. Perry, D. K. and Hannun, Y. A. (1998). *Biochim. Biophys Acta* **1436**, 233–43.
15. Spiegel, S., Foster, D., and Kolesnick, R. N. (1996). *Curr. Opin. Cell Biol.* **8**, 159–67.
16. Levade, T. and Jaffrézou, J. P. (1999). *Biochim. Biophys. Acta* **1438**, 1–17.
17. Kolesnick, R. N. and Hannun, Y. A. (1999). *TIBS* **24**, 224–5.
18. Wiegmann, K., Schütze, S., Machleidt, T. Witte, and Krönke, M. (1994). *Cell* **78**, 1005–15.
19. Liu, P. and Anderson, G. G. W. (1995). *J. Biol. Chem.* **270**, 27179–85.
20. Cifone, M. G., De Marie, R., Roncaioli, P., Rippo, M. R., Azuma, M., Lanier, L. L., Santoni, A, and Testi, R. (1995). *J. Exp. Med.* **177**, 1547–52.
21. Adam-Klages, S., Adam, D., Wiegmann, K., Struve, S., Kolanus, W., Schneider-Mergener, J., and Krönke, M. (1996). *Cell*, **86**, 937–47.
22. Schwandner, R., Wiegmann, K., Bernado, K., Kreder, D, and Krönke, M. (1998). *J. Biol. Chem.* **273**, 5916–22.
23. Wiegmann, K, Schwandner, R., Krut, O., Yeh, W.-C., Mak, T. W., and Krönke, M. (1999). *J. Biol. Chem.* **274**, 5267–5270.
24. Smyth, M. J., Perry, D. K., Zhang, J., Poirier, G. G., Hannun, Y., and Obeid, L. M. (1996). *Biochem. J.* **316**, 25–8.
25. Santana, P., Pena, L. A., Haimovitz-Friedmann, A., Martin, S., Green, D., Mc Loughlin, M., Cordon-Cardo, C., Schuchmann, E. H., Fuks, Z., and Kolesnick, R. (1996). *Cell* **86**, 189–99.
26. Herr, I., Wilhelm, D., Böhler, T., Angel, P., and Debatin, K.-M. (1997). *EMBO-J.* **16**, 6200–8.
27. De Maria, R., Rippo, M. R., Schuchmann, E. H., and Testi, R. (1998). *J. Exp. Med.* **187**, 897–902.
28. Monney, L., Olivier, R., Otter, I., Jansen, B., Poirier, G. G., and Borner, C. (1998). *Eur. J. Biochem.* **251**, 295–303.
29. Haimowitz-Friedmann, A., Cordon-Cardo, C., Bayoumy, S., Garzotto, M., McLouglin, M., Gallily, R., Edwards, C. K., III, Schuchman, E. H., Fuks, Z., and Kolesnick, R. (1997). *J. Exp. Med.* **186**, 1831–41.
30. Mathias, S., Dressler, K. A., and Kolesnick, R. N. (1991). *Proc. Natl. Acad. Sci USA* **88**, 1009–13.
31. Zhang, Y., Yao, B., Delikat, S., Bayoumy, S., Lin, X. H., Basu, S., McGinley, M., Chan-Hui, P. Y., Lichenstein, H., and Kolesnick, R. (1997). *Cell* **89**, 63–72.

32. Dobrowski, R. T., Kamibayashi, C., Mumby, M. C., and Hannun Y. (1993) *J. Biol. Chem.* **268**, 15523–30.

33. Law, B. and Rossie, S. (1995). *J. Biol. Chem.* **270**, 12808–13.

34. Chalfant, C. E., Kishikawa, K., Mumby, M. C., Kamibayashi, C., Bielawska, A., and Hannun, Y. A. (1999). *J. Biol. Chem.* **274**, 20313–20317.

35. Müller, G., Ayoub, M., Storz, P., Rennecke, J., Fabbro, D., and Pfizenmaier, K. (1995). *EMBO J.* **14**, 1961–96.

36. Huwiler, A., Brunner, J., Hummel., R., Vervoodeldonk, M., Stabel, S., van den Bosch, H., Pfeilschifter, J. (1996). *Proc. Natl. Acad. Sci. USA* **93**, 6959–63.

37. Müller, G., Storz, P., Bourteele, S., Döppler, H., and Pfizenmaier, K. (1998). *EMBO J.* **17**, 732–42.

38. Westwick, J. K., Bielawska, A. E., Dbaibo, G., Hannun, Y. A., and Brenner, D. A. (1995). *J. Biol. Chem.* **270**, 22689–92.

39. Huang, C., Ma, W., Ding, M., Bowden, G. T., and Dong, Z. (1997). *J. Biol. Chem.* **272**, 27753–7.

40. Zhang, J., Alter, N., Reed, J. C., Borner, C., Obeid, L. M., and Hannun, Y. (1996) *Proc. Natl. Acad. Sci. USA* **93**, 5325–8.

41. Schütze, S., Machleidt, T., Adam, D., Schwandner, R., Wiegmann, K., Kruse, M.-L., Heinrich, M., Wickel, M., and Krönke, M. (1999). *J. Biol. Chem.* **274**, 10203–12.

42. Mizushima, N., Koike, R., Kohsaka, H., Kushi, Y., Handa, S., Yagita, H., and Miyasaka, N. (1996). *FEBS Lett.* **395**, 267–71.

43. Anjum, R., Ali, A.M., Begum, Z., Vanaja, J., and Khar A. (1998). *FEBS Lett.* **439**, 81–4.

44. Wickel, M., Heinrich, M., Weber, T., Brunner, J., Krönke, M., and Schütze, S. (1999). *Biochem. Soc. Trans.* **27**, 393–400.

45. Heinrich, M., Wickel, M., Schneider-Brachert, W., Rosenbaum, C., Gahr, J., Schwandner, R., Weber, T., Brunner, J., Krönke, M., and Schütze, S. (1999). *EMBO J.* **18**, 5252–63.

46. Deiss, L. P., Galinka, H., Berissi, H., Cohen, O., and Kimchi, A. (1996). *EMBO J.* **15**, 3861–70.

47. Wu, S. H., Saftig, P., Peters, C., and El-Deiry, W. (1998). *Oncogene* **16**, 2177–83.

48. Shibata, M., Kanamori, S., Isahara, K., Ohsawa, Y., Konishi, A., Kametaka, S., Watanabe, T., Ebisu, S., Ishido, K., Kominami, E., and Uchiyama, Y. (1998). *Biochem. Biophys. Res. Commun.* **251**, 199–203.

49. Ohsawa, Y., Isahara, K., Kanamori, S., Shibata, M., Kametaka, S., Gotow, T., Watanabe, T., Kominami, E., and Uchiyama, Y. (1999). *Arch. Histol. Cytol.* **61**, 395–403.

50. Roberg, K. and Öllinger, K (1998). *Am. J. Pathol.* **152**, 1151–6.

51. Roberg, K., Johansson, U., and Öllinger, K. (1999). *Free Radical Biol. Med.* **27**, 1228–37.

52. Pessin, M. S., Baldassare, J. J., and Raben, D. M. (1990). *J. Biol. Chem.* **265**, 7959–66.

53. Pettitt, T. R., Zaqqa, M., and Wakelam, J. O. (1994). *Biochem. J.* **298**, 655–60.

54. Preiss, J., Loomis, C. R., Bishop, W. R., Stein, R., Niedel, J. E., and Bell, R. M. (1986). *J. Biol. Chem.* **261**, 8597–600.

55. Ha, K. S. and Exton, J. H. (1993). *J. Biol. Chem.* **268**, 10534–9.

56. Augert, G., Bocckino, S. B., Blackmore, P. F., and Exton, J. H. (1989). *J. Biol. Chem.* **264**, 21689–98.

57. Schütze, S., Berkovic, D., Tomsing, O., Unger, C., and Krönke, M. (1991). *J. Exp. Med.* **174**, 975–88.

58. Rana, R. S., Mertz, R. J., Kowluru, A., Dixon, J. F., Hokin, L. E., and MacDonald, M. J. (1985). *J. Biol. Chem.* **260**, 7861–7.

59. Bligh, V. and Dyer, W. J. (1959). *Can. Biochem. J. Physiol.* **37**, 911–17.

60. Rustenbeck, I. and Lenzen, S. (1990). *J. Chromatogr.* **525**, 85–91.

61. Huang, C. F. and Cabot, M. C. (1990). *J. Biol. Chem.* **265**, 17468–73.

62. Billah, M. M., Eckel, S., Mullmann, T. J., Egan, R. W., and Siegel, M. I. (1989). *J. Biol. Chem.* **264**, 17069–77.
63. Nakamura, K. and Handa, S. (1984). *Anal. Biochem.* **142**, 406–10.
64. Gu, M., Kerwin, J. L., Watts, J. D., and Aebersold, R. (1997). *Anal. Biochem.* **244**, 347–56.
65. Thomas, R. L., Matsko, C. M., Lotze, M. T., and Amoscato, A. A. (1999). *J. Biol. Chem.* **274**, 3580–8.
66. Perry, D. K. and Hannun, Y. A. (1999). *TIBS* **24**, 226–7.
67. Watts, J. D., Aebersold, R., Polverino, A. J., and Patterson, S. C. (1999). *TIBS* **24**, 227.

Chapter 8

Conditional and fixed cytokine gene modifications in transgenic and mutant mice

George Kassiotis*, Manolis Pasparakis, and George Kollias*

Laboratory of Molecular Genetics, Hellenic Pasteur Institute, 127 Vas. Sofias Ave, Athens 115-21, Greece

*Institute of Immunology, Biomedical Sciences Research Centre 'Al. Fleming', 14–16 Al. Fleming Str. Vari 166-72, Greece

1 Introduction

For many years, transgenic mice have served as fine tools in the dissection and understanding of complex biological phenomena. Studies on gene function in multifactorial and multicellular systems have benefited most, because they often require experimental settings which faithfully stimulate the *in vivo* situation. As additional technological advances become accessible to a growing number of investigators, these offer the opportunity to inactivate or modify endogenous genes in the mouse and provide new possibilities for research. In general, perturbation of gene expression in transgenic systems often leads to measurable alterations in the physiology of the animal and important functional implications come to light.

Within the limitations of space in this chapter, we have sought to provide a technical guide to the production of transgenic and knockout mice for use by those who have limited experience in this field, taking care to include technical tips and details that have not been published elsewhere. Detailed laboratory manuals on transgenic technologies may also be found in refs (1 and 2).

2 Production of transgenic mice by pronuclear injection of DNA

2.1 Equipment

Specialized equipment is needed for pronuclear injections. Inverted microscopes are more convenient to use for microinjection and they should be equipped with Nomarski differential interference contrast optics (DIC) or Hoffman optics, which allow better visualization of the pronuclear membranes. A pipette puller capable of generating good injection needles is an absolute requirement for successful injections. Two sets of micromanipulators are necessary for micro-injection. Two stereo microscopes and a cold light source will be needed for the surgical transfer of the injected eggs into the oviducts of the pseudopregnant females. In our laboratory we use a Diaphot TMD microscope (Nikon Ltd) with Nomarski optics and a set of Leitz micromanipulators (Leitz Instruments Ltd). A Kopf needle puller, model 750 (David Kopf Instruments), is used for needle pulling and two Nikon stereomicroscopes and a Nikon fibre-optic light source are used for the egg transfers. Other commercial suppliers of suitable equipment are Leitz, Carl Zeiss, and Olympus for the inverted and the stereomicroscopes, Narishige Co. for the micromanipulators, and Campden Instruments for the needle puller.

2.2 Animals

2.2.1 Mouse strains

Fertilized oocytes obtained from several strains of mice have been used for pronuclear injections. Zygotes produced by hybrid F_1 females are found to give better results than most inbred strains, possibly through a maternal or egg cytoplasmic effect that gives increased recovery of the eggs after microinjection. When inbred genetic backgrounds (e.g. specific MHC haplotypes, disease resistance/susceptibility) are essential, inbred instead of F_1 zygotes may be used, albeit with lower overall efficiency. At least one inbred strain, the FVB/N, shows efficiencies comparable to F_1 hybrids (3). Good F_1 females are obtained when C57BL/6 females are crossed with CBA, C3H, or SJL/J males. We use (CBA × C57BL/6) F_1 females both for the derivation of the eggs and as pseudopregnant foster mothers.

2.2.2 Scale

The levels of transgene expression may vary significantly between different mouse lines generated with the same DNA, due to the variable number of integrated copies and the unpredictable influences of neighbouring chromatin at the transgene integration site. When 'good' expression of a transgene is required the production of four to six transgenic mouse lines is usually sufficient, although in particular cases more mouse lines may be needed (e.g. when the generation of a phenotype is expected to be associated with the level of transgene expression).

To obtain four to six transgenic founders, at least 150 injected eggs should be transferred into pseudopregnant mothers. For this, 10–15 superovulated females will be needed. Depending on the supply of mice and the skills of the operator, this is usually accomplished in 2–3 days.

2.3 Preparation of DNA for microinjection

Any cloned fragment of DNA can be used for injection. Linear DNA fragments with sticky ends show much higher integration efficiencies than blunt-ended linear or supercoiled DNA (4). Prokaryotic vector sequences have been found to interfere with the expression of the transgene (5) and they should be removed before microinjection. For this reason, unique restriction sites on both sides of the insert should be considered when designing a construct. Since the presence of introns is shown to facilitate transcription of the transgene, intron-containing genes are preferred over cDNA constructs (6).

2.3.1 Purification of DNA fragments

DNA purity is a very important factor for the successful generation of transgenic mice, since impurities such as traces of agarose or organic solvents can severely decrease the viability of the injected egg. Several methods are available for isolating DNA fragments (7). Depending on the size of the construct different approaches may be followed. For large DNA fragments (> 100 kb), such as those cloned in yeast artificial chromosomes (YAC), specific successful protocols have been described (8). For medium fragment sizes, e.g. cosmid inserts or ligated cosmid inserts (9), sodium chloride gradients or preparative pulse-field gel electrophoresis may be used. For DNA fragments less than 25 kb in length, we routinely use preparative agarose gel electrophoresis followed by elution of the DNA fragment by digestion with β-agarase I (New England Biolabs) according to the manufacturer's instructions, followed by an extra purification step with tip-columns Elutip-D (Schleicher & Schuell).

2.3.2 The DNA concentration for microinjection

The DNA concentration used for microinjection is usually adjusted to 1–2 ng/μl. This concentration reflects the original descriptions referring to successful rates of integration when 500–1000 copies of a 3.5-kb linear DNA fragment were injected into each zygote (4). However, when low integration frequencies are encountered we have found that for DNA constructs ranging from 4 to 12 kb in size, the concentration can safely be increased to 4–6 ng/μl. For new DNA constructs prepared for microinjection, the concentration is adjusted by comparison to an older fragment of similar size that has been used successfully in previous experiments. The new DNA is diluted until an equal volume of solution gives equal band intensity with the control DNA on an ethidium bromide-stained agarose gel.

2.4 Preparation of the pipettes used for pronuclear injections

2.4.1 The injection pipette

Needles for microinjection are made from glass capillaries (borosilicate glass capillaries, thin wall, with an inner filament) using a pipette puller. The shape of the needle is the most important factor for egg survival after injection. With an optimal needle, up to 90% of the eggs should survive the injection. Using the

Kopf needle puller and glass capillaries purchased from Intracel Ltd, we produce good needles using the following settings:

Heat 1:	15.5
Heat 2	0
Sol:	4
Delay:	0
Sol:	0.05

The two Sol values are parameters for the pulling force and pulling time (described in the Kopf puller manual). The pulling time, which should be around 5.5 secs, can be controlled by adjusting the proximity of the heating elements.

2.4.2 The holding pipette

The holding pipette is a blunt, heat-polished pipette through which suction is applied to position and hold the egg for microinjection. Its external diameter should be between 80–120 μm with an opening of approximately 20–30 μm. It is easier to prepare optimal holding pipettes using a microforge instrument (e.g. Micro Instruments) to control the size and the shape of the opening. However, good pipettes can also be drawn and flame-polished by hand using a stereoscope and the flame of a microburner.

2.5 Recovery of oocytes

Fertilized oocytes used for microinjection are usually obtained from matings between F_1 males and females. To control the timing of ovulation it is best that animals are kept in a constant light–dark cycle. Since ovulation occurs 3–5 h after the onset of the dark period and egg pronuclei are suitable for microinjection at 15–18 h after ovulation, a convenient light–dark cycle can be worked out according to the needs of the operator. For example, in a 1900–0700 h dark, 0700–1900 h light cycle, microinjection is best performed between 1500 and 1800 h.

Since natural matings produce low numbers of zygotes it is preferable that females are induced to superovulate. For superovulation, 3–5-week-old females are injected intraperitoneally (IP) on day 1 with 5 units of pregnant mare's serum (PMS), which mimics the effects of follicle-stimulating hormone (FSH). This is done 5–6 h before the start of the dark period of the light cycle. On day 3, 46–48 h after PMS administration, the female mice are injected IP with 5 units of human chorionic gonadotropin (huCG) and placed with individually caged F_1 stud males (i.e. one female and one male per cage). To maximize the fertilizing efficiency of sperm, it is best to alternate the group of males so that each male gets one female every 2–3 days. The following morning, females are checked for vaginal plugs. Those that have mated are killed and their oviducts dissected out and placed in a 35-mm tissue culture dish containing standard M2 medium equilibrated at room temperature. Zygotes are collected in M2 medium containing 300 μg/ml hyaluronidase (Sigma) to remove the sticky cumulus cells. This step should take no longer than 10–15 min as hyaluronidase may affect egg viability. Following this, the eggs are washed several times in M2 medium to remove the hyaluronidase, transferred in CO_2-buffered M16 medium and stored at 37°C, 5% CO_2 in microdrop cultures overlaid with paraffin oil. M2 and M16 media can be prepared in the laboratory using analytical grade chemicals and pyrogen-free water according to published protocols (1, 2). We find it more convenient to buy both M2 and M16 powdered media (Sigma), and prepare them according to the manufacturer's instructions.

2.6 Microinjection procedure

Pronuclear injections are carried out in a paraffin oil-covered drop of M2 medium, set on a siliconized glass depression slide. The glass slide should be washed in a mild detergent and thoroughly rinsed in pyrogen-free water. The holding pipette, which is filled with Fluorinert FC77 (Sigma), is controlled by a micrometre syringe connected to the pipette through a length of paraffin oil-filled Tygon tubing. The injection needle is connected through air-filled tubing to an air-filled 50 ml glass syringe.

Protocol 1

Microinjection procedure

Equipment and reagents

- Injection and holding pipettes (see Sections 2.4.1, 2.4.2, and 2.6)
- Paraffin-covered drop of M2 medium (see above)
- M2 medium
- M16 medium
- Cytochalasin D

- Oocytes (see Section 2.5)
- Inverted microscope with Nomarski optics (see Section 2.1)
- One pair of micromanipulators (see Section 2.1)
- 50 ml air-filled syringe
- 5% CO_2 incubator

Method

1   First lower the holding and injection pipettes into the drop of M2 medium

2   Transfer a small number of eggs, usually 15–30 depending on the experience of the injector, into the drop.

3   Check that the injection needle is open by moving it close to one of the eggs and applying pressure while observing the movement of the egg. If the needle is closed, try to brake its tip by letting it carefully touch the holding pipette.

4   At low magnification use the holding pipette to pick an egg and observe it at higher magnification for visible pronuclei. Select the most easily accessible pronucleus and use the holding and injection pipettes to bring the egg to the best position for injection. Depending on the structure of the opening of the holding pipette, it is usually safer for the egg to be held from the area of the zona pellucida next to the polar bodies.

Protocol 1 continued

5 Move the injection needle close to the egg and use the micromanipulator adjuster to focus both the tip of the needle and the pronucleus at the same level.

6 Insert the needle through the zona pellucida and the egg membrane into the pronucleus. Apply some pressure to the 50 ml air-filled syringe—if enough pressure is applied and the needle has not yet penetrated the egg and pronuclear membranes, a 'bubble-like' structure should appear at the tip of the needle. Jab the nucleus with the needle and inject again. Look for visible swelling of the pronucleus, which indicates a successful injection, then carefully withdraw the needle (*Figure 1*). Note that after injection the pronucleus tends to shrink back to its original size.

7 Move the injection oocyte to one side of the optical field using low magnification and proceed to the next one.

8 After having injected the first group of eggs, transfer them into M16 medium and store in the CO_2 incubator. Continue with the rest of the groups.

An experienced injector should expect that approximately 70–90% of the eggs will survive injection. However, ease of injections and survival rates may increase when cytochalasin D (at 1 μg/ml) is included in the drop of M2 medium where injections are taking place. Cytochalasin D reversibly depolymerizes the cytoskeleton of the egg, making the cell membranes more distortable and less susceptible to lysis. Injected eggs should be washed extensively in M2 medium to remove cytochalasin D.

Protocol 2

Egg transfer into the oviducts of pseudopregnant females

Equipment and reagents

- Anaesthetic (Hypnorm/Hypnovel/double-distilled H_2O mix 1:1:6)
- Blunt and fine forceps
- Surgical wound clips (e.g. disposable skin stapler, Precise DS-25 from 3M)
- Binocular dissecting microscope and cold light source
- 0.1% adrenaline
- Mouth-driven, glass transfer pipette
- Heat source for post-anaesthetic recovery

Mature (> 7 week's old) F_1 females, mated with vasectomized males are used as pseudopregnant recipients for the injected eggs. Egg transfer may take place on either the same day of injection or the next day, when the eggs will be at the two-cell stage. In either case, day 0.5 pseudopregnant females should be used as recipients.

Method

1 Anaesthetize the recipient female by intraperitoneal injection of 0.3 ml of Hypnorm/Hypnovel/ddH_2O mix (10).

2 Swab the back of the mouse with ethanol and make a small midline incision in the skin at approximately the level of the last rib. Locate the position of the ovary indicated by a pink structure seen through the body wall.

Protocol 2 continued

3 Make a small incision through the body wall and use blunt forceps to grasp and gently pull out the fat pad which is attached to the ovary, thus exposing the uterus.

4 Place a clip on the fat pad to hold the oviduct in place and move the mouse to under a binocular dissecting microscope.

5 Expose the opening of the oviduct, which is found adjacent to the ovary, by tearing the bursa with fine forceps. Rupture of the bursa often causes bleeding which can obscure vision and lead to operational difficulties for the inexperienced. To avoid this, apply a drop of a 0.1% solution of adrenaline directly on to the bursa just before tearing it. Note that this approach has no apparent adverse effects and prevents all bleeding from the ruptured bursa (G. Schmidt and J. O'Sullivan, personal communication).

6 Expel the eggs into the oviduct using a mouth-driven, glass transfer pipette. Transfer approximately 15 eggs at the one-cell stage (or 10 eggs if at the two-cell stage) to each oviduct.

7 When the operation is completed, staple the edges of the skin together where the incision was made using one or two wound clips.

It is important to keep the mouse warm after the surgery until it has recovered from the anaesthetic. Two or three recipient females are caged together and they usually give birth 19–20 days after the transfer. Caging the females together lowers the possibility of losing some newborns due to small litter size.

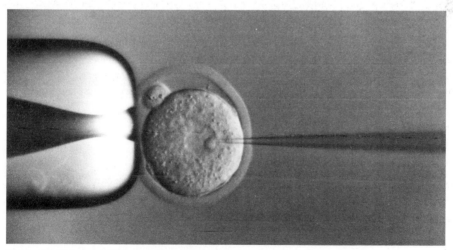

Figure 1 Injection of DNA solution into one of the pronuclei of a mouse zygote.

2.7 Integration and expression of injected genes

A few hundred copies of the gene of interest are usually injected into the pronucleus of the mouse zygote. The factors affecting subsequent integration into the mouse genome have not yet been defined. In general, it is accepted that integration occurs at a random site in the mouse genome and that transgenes are usually found inserted at a single chromosomal locus, either as single copies or, more often, in tandem, head-to-tail arrays which are inherited as Mendelian traits. More rarely, transgene integration can occur at two or more positions in the mouse genome. In addition, it has been observed that in as many as 30–40% of the transgenic founder mice only a subset of cells carry the injected DNA, and these mice are therefore mosaic for the transgene. Such mosaicism could result from delayed DNA integration at the two-cell or even later stages of development. Mosaicism is not always a problem, since it can contribute to the survival of mice carrying an otherwise lethal or pathology-inducing transgene. This may be a common situation in transgenic studies of cytokine function. For example, we have been able to analyse the pathology induced by human TNF in transgenic mice by studying the development of an early-lethal phenotype in the progeny of an unaffected mosaic transgenic founder (11).

The transcriptional efficiency of transgenes is almost always influenced by the activation status of the neighbouring chromatin at the insertion site. It is generally observed that both the specificity and the levels of transgene expression may be influenced. However, in most cases, it has been shown that tissue specificity can be correctly conferred upon expression of the transgene, by including a few hundred base pairs of flanking, *cis*-acting DNA sequences (e.g. promoters and enhancers) into the gene construct. Influences from the neighbouring chromatin do not affect the expression of such transgenes in the 'correct' tissues, although they can often cause expression in 'wrong' tissues. At present, position-independent expression of transgenes can only be obtained using specific *cis*-acting DNA elements called locus control regions (LCRs) which are able to drive position-independent, tissue-specific, and copy number-dependent expression of the associated transgenes (12, 13).

Protocol 3

Preparation of DNA from tail fragments of mice

Equipment and reagents

- Tail buffer: 50 mM Tris–HCl pH 8.0, 0.1 M EDTA, 0.1 M NaCl, 1% SDS
- 10 mg/ml proteinase K in 50 mM Tris–HCl buffer pH 8.0
- 25:24:1 (v/v) phenol/chloroform/isoamyl alcohol saturated in 100 mM Tris–HCl buffer pH 8.)
- 10 mg/ml RNase A in 0.3 M NaCl, 0.03 M Na_3-citrate. Inactivate DNase impurities by a 5-min incubation in a boiling waterbath
- 1.5 ml safe-lock polypropylene tubes (e.g. Eppendorf)
- Vortex mixer
- 24:1 (v/v) chloroform/isoamyl alcohol
- Isopropanol
- 70% ethanol
- TE buffer: 10 mM Tris–HCl pH 8.0, 1 mM EDTA
- Microcentrifuge
- Pipette tips with cut ends
- Sealed Pasteur pipettes
- Bench-top shaker
- UV spectrophotometer

Protocol 3 continued

Method

1 Cut off 0.5–1 cm of the tail into a 1.5 ml safe-lock polypropylene tube containing 0.6 ml of tail buffer.

2 Add 20 μl of 10 mg/ml proteinase K solution, mix and incubate overnight at 55 °C.

3 The next morning add 1 μl of 10 mg/ml RNase A and incubate for 1 h at 37 °C.

4 Add 0.6 ml of the phenol/chloroform/isoamyl alcohol to each tube and vortex for 5–10 min.

5 Centrifuge each tube for 10 min at 13 000 g in a microcentrifuge at room temperature. Remove the supernatant to a fresh tube using pipette tips with cut ends to avoid any carryovers from the interface.[a]

6 Repeat steps 4 and 5.

7 Add 0.5 ml of chloroform/isoamyl alcohol and vortex for 5 min.

8 Centrifuge for 5 min at 13 000 g in a microcentrifuge and remove the aqueous phase to a new tube.

9 Proceed with one tube at a time. Add 0.6 vol. of isopropanol to each tube, mix several times by inversion until a white DNA mass is visible. Hook out the precipitated DNA using a sealed Pasteur pipette.

10 Immerse the DNA briefly in 70% ethanol, let it dry for a few minutes, and place it in a microcentrifuge tube containing 100 μl of TE buffer. Allow to stand for 10 min and then carefully remove and discard the Pasteur pipette.

11 Shake the tubes on a bench-top shaker to completely dissolve the DNA, then determine the concentration by measuring the A_{260} OD.

[a] The quality of the DNA preparation at this stage is appropriate for slot-blot hybridization analysis (see Protocol 4). It can also be stored at −20 °C for later further purification. However, if the DNA is to be used for Southern hybridization analysis, the protocol should be carefully followed to the end.

Protocol 4

Slot-blot analysis of tail DNA

Equipment and reagents

- Slot-blot apparatus (e.g. Schleicher & Schuel)
- Tail DNA: 5 μg of DNA or 20 μl of the supernatant from the first phenol extraction of tail DNA (see Protocol 3, step 5)
- X-ray film
- Hybridization equipment and reagents
- Positively charged nylon membrane (e.g. Hybond N$^+$ from Amersham)

- Alkaline solution: 0.5 M NaOH, 1.5 M NaCl
- Neutralizing solution: 0.5 M Tris–HCl pH 7.4, 1.5 M NaCl
- Radiolabelled probe for endogenous single-copy gene
- Radiolabelled probe for transgene detection
- 4 M NaOH

Protocol 4 continued

Method

1   For each sample, add 5 μg of DNA (or 20 μl of phenol supernatant) in water to a final volume of 180 μl. Then add 20 μl of 4 M NaOH, mix and allow to stand at room temperature for 15 min.

2   Cut a piece of nylon membrane to the correct size, rinse in distilled water, and soak in alkaline solution for at least 10 min.

3   Assemble the slot-blot apparatus according to the manufacturer's instructions and wash each well twice with alkaline solution. Apply the samples.

4   Disassemble the apparatus, immerse the membrane in 200 ml of neutralizing solution, and shake gently for 10 min at room temperature.

5   Air-dry the membrane and cut it so that each well is divided into two parts of equal size (see *Figure 2*).

6   Hybridize half the membrane with a radiolabelled probe for an endogenous single-copy gene and the other half with the appropriate radiolabelled prove for transgene detection. This approach efficiently controls for false-negatives or false-positives and minimizes general quantitative artefacts. It is recommended especially when low-copy transgenic lines are analysed.

7   After hybridization assemble the pieces of the membrane and expose on X-ray film.

2.8 Identification of transgenic progeny

Identification of transgenic mice is generally performed by standard Southern or dot/slot blot hybridization techniques on genomic DNA prepared from mouse tail fragments. Alternatively, transgene-specific PCR amplification may also be applied, which allows the use of smaller samples of tissue, such as tail tips, ear pieces, or blood samples.

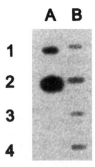

Figure 2 Slot-blot analysis of mouse-tail DNA. Each slot on the membrane is cut into two parts and each piece hybridized with a different probe. In this example, the left half (a) is hybridized with a radiolabelled probe specific for the detection of human p55 TNF-receptor transgenes, and the right half (b) with a radiolabelled prove specific for the single-copy endogenous mouse p55 TNF-receptor gene. The latter is used as a quantitative control.

    Slot 1:   Tail DNA from a low-copy number transgenic mouse.
    Slot 2:   Tail DNA from a high-copy number transgenic mouse.
    Slot 3:   Tail DNA from a non-transgenic littermate.
    Slot 4:   Normal mouse DNA.

Protocol 5

Screening transgenic animals using PCR

Equipment and reagents

- PCR lysis buffer: 50 mM KCl, 10 mM Tris–HCl pH 8.3, 1.5 mM MgCl$_2$, 0.1 mg/ml gelatin, 0.45% IGEPAL CA-630 (Sigma), 0.45% Tween-20
- 10 × PCR buffer: 670 mM Tris–HCl pH 8.8, 166 mM (NH$_4$)$_2$SO$_4$, 1 mg/ml BSA
- Waterbaths at 56 °C and 95 °C
- Microcentrifuge and tubes
- 2000 U/ml *Taq* polymerase
- Sterile distilled water
- Paraffinoil (depending on the PCR apparatus used)

- 10 mg/ml proteinase K
- 10 mM MgCl$_2$
- dNTP mixture (2.5 mM each)
- Specific oligonucleotides (sense and antisense) for the transgene
- 10 × TBE buffer: 0.89 M Tris base, 0.89 M boric acid, 25 mM EDTA. Adjust to pH 8.3
- Loading buffer
- 1.5% agarose gel in 1 × TBE buffer
- UV transilluminator

Method

1 Cut a small piece (not more than 2 mm) from the tip of the tail and digest overnight at 56 °C in 250 μl of PCR-lysis buffer containing 3 μl of 10 mg/ml proteinase K solution.

2 The next day incubate the tail digests at 95 °C for 30 min to inactivate the proteinase.

3 Cool on ice for 10 min, spin at full speed for 10 min in a microcentrifuge, and remove 2 μl from each sample into the tubes for PCR.

4 Prepare enough reaction mixture for all samples, each reaction requires:

2 μl 10 × PCR buffer

2 μl 10 mM MgCl$_2$

1.5 μl dNTP mixture (2.5 mM each)

1 μl oligonucleotide mix[a]

1 μl *Taq* polymerase (2000 U/ml)

10.5 μl sterile distilled water

5 Add 18 Xml of the reaction mixture to each tube, overlay each reaction with paraffinoil (depending on the model of the PCR apparatus used), and cycle the tubes using standard appropriate conditions.

6 When cycling is completed, add loading buffer to each reaction, and run the samples on a 1.5% agarose gel in 1 × TBE buffer.

7 Visualize the bands on a UV transilluminator and photograph the gel.

[a] The optimal concentration for each pair of oligonucleotides should be established by carrying out reactions with concentrations ranging from 0.5 to 10 pmol of each oligonucleotide per reaction.

3 Production of knockout mice

The embryonic stem-cell (ESC) technology, combined with the methodology of gene targeting by homologous recombination, has made it possible to introduce designed mutations into the mouse germline. With the use of this technology, any desired genetic alteration can be introduced into the genome of ESC in culture, which is then transferred into the mouse germline, making it possible to study the function of the mutated gene in the animal.

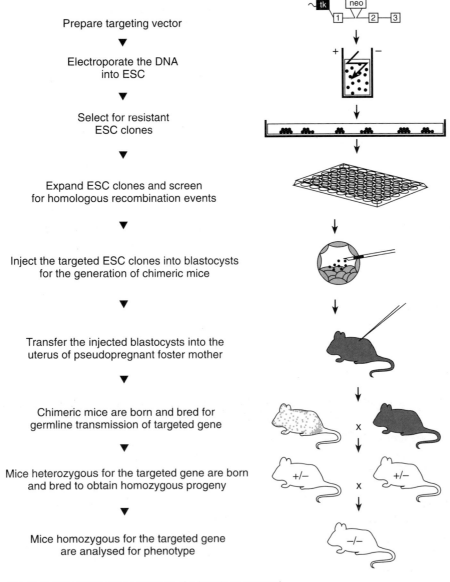

Prepare targeting vector

▼

Electroporate the DNA
into ESC

▼

Select for resistant
ESC clones

▼

Expand ESC clones and screen
for homologous recombination events

▼

Inject the targeted ESC clones into blastocysts
for the generation of chimeric mice

▼

Transfer the injected blastocysts into the
uterus of pseudopregnant foster mother

▼

Chimeric mice are born and bred for
germline transmission of targeted gene

▼

Mice heterozygous for the targeted gene are born
and bred to obtain homozygous progeny

▼

Mice homozygous for the targeted gene
are analysed for phenotype

Figure 3 Schematic representation of a knockout protocol.

3.1 Embryonic stem cells (ESC)

Embryonic stem cells are derived directly from the inner cell mass of mouse blastocysts, and when cultured *in vitro* they retain their undifferentiated pluri-potential character even after a significant number of passages (14). The most useful property of ESC has been their ability to contribute to the development of all cell lineages, including the germ cells, when transferred back into the mouse embryo (15). This feature makes it possible to generate mice derived ex-clusively from a single cell that has been cultured and manipulated *in vitro*. ESC lines have been derived from both inbred and outbred strains of mice. However, most existing ESC lines are derived mainly from the 129/J and 129/Sv strains, although the C57BL/6 mouse strain has also been used.

3.2 Culturing ESC

An extensive description of the methods for isolating and culturing ESC is beyond the scope of this chapter, and those who require a detailed laboratory manual for these methods are referred to refs 2 and 16.

Optimal culture conditions are critical to the *in vitro* maintenance of ESC. Good care should be given to the quality of the culture medium and serum. Since only a limited number of sera are found to be of sufficient quality for ESC, fetal calf serum should be routinely batch-tested for its ability to support their growth. The ability of ESC to remain pluripotential in culture is shown to be dependent mainly on the presence in the culture medium of the cytokine leukaemia inhibitory factor (LIF) (17, 18). Even though recombinant LIF can be supplemented in the ESC medium, the most effective and economical approach for maintaining pluripotential ESC is to culture them on layers of mitotically inactivated feeder cells that produce and secrete LIF and other factors into the medium.

3.2.1 Preparation of feeder-cell layers

Both permanently growing cell lines (i.e. STO fibroblasts) and primary mouse-embryo fibroblasts (PMEFs) have been successfully used as feeders for the main-tenance of pluripotential ESC. We use STO (neo[r]) fibroblasts grown routinely in DMEM plus 7% FCS, for the preparation of feeder cells.

Protocol 6

Preparation of PMEFs[a]

Equipment and reagents

- 13-day pregnant mice
- Fine forceps
- 3 ml syringe with 18-gauge needle
- Ca^{2+}/Mg^{2+}-free PBS
- DMEM plus 7% FCS
- 10 cm tissue culture dish
- Trypsin–EDTA solution: 1:250 trypsin 0.2 g/litre EDTA, in Modified Puck's saline (Gibco-BRL)
- 37°C, 5% CO_2 incubator
- Tissue culture centrifuge

Protocol 6 continued

Method

1 Set up females with males (any mouse strain can be used for the preparation of PMEFs but a neo^r-carrying strain is preferred). The following day check for copulation plugs (this is day 1 of pregnancy). Remove the mated females to a separate cage.

2 On day 13 of pregnancy, kill the females, open the peritoneal cavity and remove the uteri carrying the embryos. Dissect the embryos from the uteri in PBS and remove the yolk sac, amnion, and placenta. Wash the embryos twice in fresh PBS to remove any blood.

3 Using a pair of fine forceps pinch off the head and pinch out and remove the liver.

4 Place 5–8 embryos in the barrel of a 3-ml syringe to which a sterile 18-gauge needle is attached.

5 Add 2 ml of sterile PBS and replace the syringe plunger. With the tip of the needle in a 10 cm tissue culture dish, expel and draw up the embryos through the needle 4–5 times, breaking them into small clumps of cells.

6 Add 20 ml of DMEM plus 7% FCS split into two 10 cm dishes and place them in the incubator. Note that the clumps will start to give rise to fibroblasts over the next 2–3 days, by the third day the dishes should be confluent.

7 Was the dishes three times with 5 ml of PBS. Trypsinize the cells with 1 ml of the trypsin–EDTA solution and collect them in DMEM plus 7% FCS.

8 Centrifuge at 250 g for 5 min at RT and resuspend the cell pellet in DMEM plus 7% FCS. Transfer each dish to three other 10 cm dishes. After further culture the fibroblasts can be split at no more than 1:3 ratio to expand their numbers. However, PMEFs are only good as a feeder layer for five or six passages, since they then will begin to cease proliferating and fresh cultures will need to be established.

9 Once sufficient numbers of PMEFs are established. either prepare them as a feeder layer for ESC, or store frozen in liquid nitrogen.

^a Adapted from ref. 2.

Protocol 7

Preparation of feeder cell layers

Equipment and reagents

- DMEM plus 7% FCS
- Ca^{2+}/Mg^{2+}-free PBS
- Trypsin–EDTA solution: 1:250 trypsin, 0.2 g/litre EDTA, prepared in Modified Puck's saline (Gibco-BRL)
- 10 cm tissue culture dish
- 37 °C, 5% CO_2 incubator
- Tissue culture centrifuge

- 1 mg/ml mitomycin C (Sigma) in PBS. Aliquot and store at −70 °C. Thaw an aliquot just prior to use.
- 0.1% gelatin solution. Prepare by diluting a 2% gelatin solution (Sigma cell-culture reagents, type B, from bovine skin) in sterile double-distilled H_2O.

Protocol 7 continued

Method

1 To a confluent 10 cm dish of STO fibroblasts add mitomycin C at a final concentration of 10 μm/ml and incubate at 37°C, 5% CO_2 for 2–3 h.

2 To prepare gelatinized tissue culture dishes add enough 0.1% gelatin solution to cover the surface of the dishes. Incubate at room temperature for at least 1 hour. Aspirate the gelatin solution before plating the feeder cells.

3 Aspirate the mitomycin C-containing medium from the cells and wash them three times with 5 ml of PBS. Trypsinize the cells with 1 ml of the trypsin–EDTA solution and collect them in DMEM plus 7% FCS.

4 Centrifuge at 250 g for 5 min at RT and resuspend the cell pellet in DMEM plus 7% FCS. Count the cells and dispense the cell suspension in gelatin-coated dishes at a concentration of 6–7 × 10^4 cells/cm^2.

5 Note that feeder cells should spread to give a monolayer in a few hours and if fed with fresh medium weekly, they can be used for as long as 15–20 days after their preparation. Change the medium to ESC medium before adding ESC.

3.2.2 The ESC culture medium

The medium used is Dulbecco's modified Eagle's medium (DMEM) with high glucose (4.5 g/litre), L-glutamine, non-essential amino acids, and no sodium pyruvate. DMEM and phosphate-buffered saline (Ca^{2+}/Mg^{2+}-free PBS) is purchased in 1 × liquid form, to avoid any variations in the quality of water used in its preparation. L-glutamine is added to a final concentration of 2 mM, penicillin (100 U/ml) and streptomycin (100 μl/ml) are included, and 2-merceptoethanol is added to a final concentration of 0.1 mM. DMEM and PBS in 1 × liquid form and L-glutamine, MEM non-essential amino acids, and penicillin/streptomycin in 100 × solutions are obtained from Gibco-BRL. The 100 × solution of 2-mercaptoethanol is prepared by diluting 7 μl of a standard 14 M solution (Sigma) into 10 ml of PBS. The medium is supplemented with 15% fetal calf serum and 10^3 U/ml LIF; this formulation is referred to as ESC medium.

3.3 Gene targeting in ESC

Gene targeting by homologous recombination in mammalian cells is a very infrequent process. The ratio of homologous recombination is found to vary significantly between different experiments. In recent reports, increased homologous integration frequencies have been achieved using DNA sequences isogenic to the target ESC DNA sequences (19).

3.3.1 Vectors for gene targeting

A targeting vector is designed to recombine with and mutate a specific chromosomal locus. For this, a vector should contain DNA sequences homologous to the

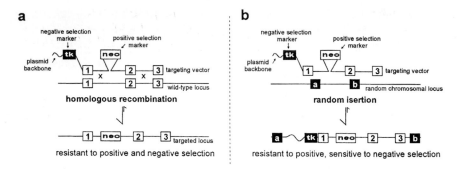

Figure 4 Schematic representation of vector/target recombination events using the positive–negative selection approach. (a) Gene targeting via homologous recombination, (b) Random vector insertion and the ESC genome.

target site and selection markers to enrich for the rare homologous recombination event.

Since homologous recombination events are highly infrequent, it is desirable to apply efficient selection schemes to enrich for the targeted clones. A positive–negative selection approach, devised by Mansour *et al.* (20), has been successfully used for many different targeting experiments. Positive selection, usually conferred by a neomycin-resistance gene cassette (*neo*), serves as a transfection marker, and, in many cases, as a means of disrupting or replacing coding exons and therefore inactivating genes. Negative selection usually imposed by the herpes simplex virus (HSV) thymidine kinase (*tk*) gene product, is useful for eliminating cells in which random integration has occurred (see *Figure 4*).

3.3.2 Strategies for the introduction of subtle mutations into the ESC genome

Several methods have been developed to introduce subtle mutations into the ESC genome. These include: microinjection of the targeting construct into ESC followed by PCR screening to identify the positive clones (21; co-electroporation of a targeting construct with an unlinked selectable marker into ESC (22); and the two-step 'in–out' or 'hit-and-run' method developed by Valancious and Smities 1991 (23) and Hasty *et al.* (24).

The 'double replacement' approach has also been developed for introducing subtle mutations into ESC (25). In this approach, endogenous sequences from the gene of interest are first exchanged by replacement with a positive and a negative selection marker (*neo* and *tk*, respectively). In a second step, the introduced *neo/tk* cassette is replaced with sequences containing the desired mutation using a new targeting construct. Targeted clones are then selected for the absence of the negative selection marker. The advantage of this method is that a number of different mutations can be efficiently introduced into the same chromosomal locus by homologous targeting in the engineered *neo/tk*-containing ESC clone.

Using a different approach, conditional or cell-type specific activation or inactivation of gene expression can be achieved (see Section 4).

3.4 Transfection of ESC

Several methods are available for the introduction of DNA into mammalian cells. Amongst them electroporation is the most widely used for gene targeting in ESC.

Protocol 8

Electroporation of DNA into ESC

Equipment and reagents

- Targeting DNA construct purified on a CsCl gradient (Qiagen prep. also works very well) and linearized with the appropriate restriction enzyme
- ES-medium (see Section 3.2.2)
- Trypsin–EDTA solution: 1:250 trypsin, 0.2 g/litre EDTA, prepared in Modified Puck's saline (Gibco-BRL)
- Ca^{2+}/Mg^{2+}-free PBS
- Plastic tissue culture dishes (10 ×O 2 cm)
- Gel electrophoresis equipment and reagents
- G418 (Genetician, Gibco)

- Electroporation apparatus: Bio-Rad Gene Pulser with a capacitance extender (Bio-Rad)
- Bio-Rad electroporation cuvettes (0.4 cm gap, Bio-Rad)
- Ganciclovir (SYMEVENE)
- Phenol/chloroform/isoamyl alcohol (see *Protocol 3*)
- Chloroform/isoamyl alcohol (see *Protocol 3*)
- Ethanol
- Sterile double-distilled water
- Tissue culture centrifuge
- Sterile, 1.5 ml tubes

Method

1 Prepare the targeting vector DNA on a CsCl gradient and linearize it by digestion with the appropriate restriction enzyme. Check by gel electrophoresis that digestion is complete and extract the DNA by treatment with phenol/chloroform and chloroform. Precipitate with ethanol, wash with 70% ethanol, and resuspend in sterile ddH$_2$O to 0.5 μg/μl.

2 One day before electroporation, passage subconfluent cultures of ESC 1:3. Change the medium 3–4 h before harvesting them on the day of electroporation to ensure that the cells are actively growing.

3 Wash the plates twice with 10 ml PBS, add 2 ml of the trypsin–EDTA solution and place them back into the incubator for 4–5 min. Add 4 ml of medium to each plate and pipette vigorously to achieve a near single-cell suspension.

4 Centrifuge the cells at 250 g for 5 min at RT and resuspend them in PBS (10 ml for every 10 cm plate). Count cell numbers.

5 Re-centrifuge the cells and resuspend them in PBS at a final concentration of 1.25×10^7 cells/ml.

6 Mix 0.8 ml of the cell suspension with 25 μg/50 μl) of the targeting vector in a sterile 1.5 ml tube and let stand for 5 min at room temp.

Protocol 8 continued

7 Transfer the suspension into an electroporation cuvette, place the cuvette in the electroporation chamber and apply a single pulse at 230 V, 500 μF. Tap the cuvette on the bench to suspend the cells and incubate for 5 min at room temp,

8 Plate the contents of each cuvette in three to five 10 cm plates containing feeder cells freshly fed with ESC medium.

9 At 24 h after electroporation apply G418 selection (use the minimum concentration of F418 that can kill all untransfected cells within 5–6 days as determine by titration for each specific ESC line; usually 150–200 μg/ml active concentration). If negative selection is required apply 2×10^{-6} M ganciclovir on days 5–8 post-electroporation.

10 Feed the cells with fresh medium every day. Note that resistant colonies should be visible 7–8 days after electroporation and ready to be picked 2–3 days later.

3.5 Picking and expanding ESC colonies

ESC colonies are ready to be picked approximately 10 days after electroporation. Single colonies are picked and transferred into single wells with feeder cells in 96-well plates. When the clones have grown to subconfluency they are split into replica plates for freezing and for DNA isolation.

Protocol 9

Picking and expanding ESC colonies

Equipment and reagents

- ESC medium (see Section 3.2.2)
- Trypsin–EDTA solution: 1:250 trypsin, 0.2 g/litre EDTA, prepared in Modified Puck's saline (Gibco-BRL)
- Tissue culture 96-well plates (with flat-bottom and -shaped wells)
- 8-channel multichannel pipette (CAPP, Denmark)

- Ca^{2+}/Mg^{2+}-free PBS
- Multipette with 8-channel adapter for dispensing liquid media into wells (Eppendorf)
- 8-place manifold aspirator for 96-well plates (Drummont Scientific Co)
- Sterile, disposable pipette tips

Method

1 Add 25 μl of trypsin–EDTA solution in each well of a U-shaped 96-well plate.

2 Wash the plate containing the ESC colonies with 10 ml of PBS and fill it with 10 ml of PBS.

3 Use a stereomicroscope, a 20 μl micropipettor, and sterile disposable pipette tips to pick individual colonies into a small volume of PBS (5–10 μl). Transfer individual colonies to the plate prepared in step 1.

Protocol 9 continued

4 After having finished with one 96-well plate, place it in the incubator for 3–5 min.

5 Take a flat-bottomed 96-well plate containing feeder cells, aspirate the medium, and add 100 μl of ESC medium to each well.

6 Remove the plate containing the trypsinized colonies from the incubator and add 100 μl of ESC medium to each well. Use a multichannel pipettor (8–12 channel) and sterile disposable pipette tips (change tips for each set of wells) to dissociate the ESC colonies by vigorously pipetting them up and down.

7 Transfer the contents of each well to the respective wells of the plate prepared in step 5.

8 Let the cells grow for the next 2–3 days, change the medium every day.

9 When cells are approaching confluency, split them 1:3. For this, wash each well twice with 100 μl of PBS, add 50 μl of trypsin–EDTA solution and let them stand in the incubator for 3–5 min. Add 100 μl of ESC medium into each well, dissociate the colonies by vigorous pipetting and split the contents in two feeder cell-containing 96-well plates. (From these plates, two will be frozen at −70°C and one will be split in three plates for DNA analysis.)

10 Freeze one of the plates 24–48 h after the split, when the fastest growing clones are approaching confluency (see *Protocol 10*). 24 h later, when the slowest growing clones are approaching confluency, freeze the second plate and split the third plate into three gelatinized 96-well plates without feeders—this will be used to prepare DNA (see *Protocol 12*).

Protocol 10

Freezing ESC clones in 96-well plates

Equipment and reagents

- Trypsin–EDTA solution: 1:250 trypsin, 0.2 g/litre EDTA, prepared in Modified Puck's saline, (Gibco-BRL)
- Ca^{2+}/Mg^{2+}-free PBS
- 2 × freezing medium: 20% DMSO, 80% FCS
- Sterile, light paraffin oil (Sigma)
- 96-well plates containing feeder cells
- ESC clones (Section 3.6) and ESC medium (Section 3.2.2)

- 8-channel multichannel pipette (CAPP, Denmark)
- Multipette with 8-channel adapter for dispensing liquid media into wells (Eppendorf)
- 8-place manifold aspirator for 96-well plates (Drummont Scientific Co)
- Parafilm
- Styrofoam box

Method

1 Grow ESC clones to subconfluence in feeder cell-containing 96-well plates.

2 Feed the cells with fresh medium 2–4 h before freezing.

Protocol 10 continued

3 Aspirate the medium and wash the wells twice with 100 μl of PBS.

4 Add 50 μl of trypsin solution per well and place the plates back into the incubator for 5–10 min.

5 Remove the plate from the incubator and add 50 μl of 2 × freezing medium in each well. Pipette up and down using the multichannel pipette (avoid bubbling) until the ESC clumps are dispersed into a near single-cell suspension.

6 Add 100 μl of sterile, light paraffin oil to each well, to prevent evaporation during storage at −70 °C. Seal the 96-well plate with Parafilm, place in the Styrofoam box, and store at −70 °C until analysis of DNA is completed.

3.6 Storage and recovery of ESC clones

Screening large numbers of ESC clones may be a time-consuming process. To minimize the time that ESC are kept in culture, it is preferable to freeze them down until screening is completed. ESC may be frozen directly in the 96-well plate making it possible to simultaneously freeze a large number of clones.

Protocol 11

Thawing ESC clones from 96-well plates

Equipment and reagents

- ESC medium (see Section 3.2.2)
- Sterile distilled water
- Sterilized Pyrex dish
- Tissue culture 24- or 48-well plates containing feeder cells
- Laminar-flow hood

Method

1 Prepare in advance 24- or 48-well plates containing feeder cells. Before use, feed with 2 ml of fresh ESC medium.

2 Warm sterile distilled water to 38–40 °C and pour into a sterile Pyrex dish, use the laminar-flow hood.

3 Remove one frozen 96-well plate from the −70 °C freezer and place it directly on the surface of the water, take care to prevent water entering the wells.

4 Hold the plate until its contents are thawed. Transfer each of the selected clones into a sterile 15 ml tube containing 2 ml of ESC medium. Wash the well with 200 μl of ESC medium to make sure that all ESC are transferred into the 15 ml tube. Spin at 250 g for 5 min, discard the supernatant. Resuspend the cells in 0.5–1 ml of ESC medium and transfer the thawed clones into the individual wells of the plate prepared in step 1.

5 Feed the cells every day and passage when they become subconfluent.

3.7 Identification of targeted ESC clones

Screening for targeted ESC clones can be performed either by PCR or by Southern blot hybridization. PCR is often used for screening pooled clones, and the individual clones of the positive pools are further analysed by Southern blotting. To minimize the risk of false-negative clones we prefer to directly analyse clones by Southern blotting and hybridization.

Protocol 12

Extraction and restriction enzyme digestion of DNA in 96-well plates[a]

Equipment and reagents

- Ca^{2+}/Mg^{2+}-free PBS
- Lysis buffer: 10- mM NaCl, 10 mM Tris–HCl pH 7.5, 10 mM EDTA, 0.5 Sarcosyl, 0.4–1 mg/ml freshly added proteinase K
- Gelatinized 96-well plates (see Protocol 7, step 2)
- ESC clones
- 70 and 100% ethanol

- Box containing wet paper towels, pre-warmed to 56°C
- Restriction digest mix: 1 × restriction enzyme panel buffer, 1 mM spermidine, 1 mM DTT, 100 μg/ml BSA, 50 μg/ml RNase A, 20 U of restriction enzyme per reaction
- 56°C incubator
- Low-power microscope

Method

1 Grow ESC clones in gelatinized 96-well plates until fully confluent.

2 Wash each well twice with 100 μl of PBS and add 50 μl of lysis buffer.

3 Transfer the plate into a 56°C pre-warmed box containing wet paper towels to create a humidified atmosphere and incubate in a 56°C oven overnight.

4 Allow the box to cool at room temperature for 1 h.

5 Add 100 μl of 100% ethanol into each well and let the plate stand on the bench for 1–2 h. Check for a filamentous DNA precipitate which should be visible under low-power magnification.

6 Invert the plate carefully to discard its contents and drain on paper towels. Note that most of the DNA should remain attached to the bottom of the wells.

7 Wash the wells three times with 100 μl of 70% ethanol, discarding each wash by carefully inverting the plate.

8 After the last wash, air-dry the DNA-containing plate on the bench. Do not let the DNA dry completely as it will be then difficult to dissolve.

9 Prepare the restriction digestion mix.

10 Add 35 μl of the digestion mix into each well and incubate overnight at the appropriate temperature in a humidified atmosphere.

Protocol 12 continued

11 The next day load the digested DNA samples on agarose gels and prepare for Southern hybridization analysis.

^a Adapted from ref. 26.

3.8 Generation of chimeric mice

Once targeted ESC clones have been identified, chimeric mice can be generated by one of several methods, including the injection of ESC into blastocysts, aggregation or co-culture of 8-cell stage embryos with ESC, and injection of ESC into 8-cell stage embryos. Methods for the aggregation (27) and co-culture (28) of 8-cell stage embryos with ESC are relatively simple and they do not require the sophisticated equipment used in the microinjection procedures. However, as yet, in many laboratories blastocyst injection is preferred over these methods as they seem to require tricky setting-up procedures. Morula injection has been developed relatively recently (27) as an alternative to the blastocyst injection method. Using this method three to six ESC are injected under the zona pellucida of 8-cell embryos and placed adjacent to the blastomeres. Initial results obtained with this method have shown lower embryo implantation frequencies in comparison to the blastocyst injection method. However, the extremely high degree of chimerism obtained in those embryos that finally implant and develop makes the morula injection method a good alternative to that of blastocyst injection.

Blastocyst injection is currently the most widely used method for the production of germline chimeras using targeted ESC clones. For ESC lines derived from the 129 strain of mice, C57BL/6 blastocysts are commonly used as hosts since they have been shown to produce the best yields of germline chimeras (29)

Protocol 13

Production of germline chimeras by injection of ESC into mouse blastocysts

Equipment and reagents

- ESC medium (see Section 3.2.2)
- Hepes-buffered ESC medium: ESC medium containing 20 mM Hepes pH 7.4
- Trypsin–EDTA solution:1:250 trypsin, 0.2 g/litre EDTA, prepared in Modified Puck's saline (Gibco-BRL)
- Ca^{2+}/Mg^{2+}-free PBS
- 1 ml syringes with 25-cauge needles
- Glass Pasteur pipettes

- Tissue culture dishes (3.5 and 6 cm)
- Borosilicate glass capillaries, thin wall without inner filament (Intracel Ltd)
- Micrometer head, 0–25 mm, 0.5 μm (Mitutoyo Co, available from Pillar Engineering Supplies Ltd)
- Microforge (Micro Instruments ltd)
- Anaesthetic: Hypnorn/Hypnovel/ddH_2O mix 1:1:6)

- C57BL/6 adult males and females
- F_1 (CBA x3 C57BL/6) adult females and vasectomized males
- Sterile dissecting instruments, including blunt forceps
- Wound clips
- Stereomicroscope
- Phase-contrast optics, 200 × magnification
- Microdrop cultures overlaid with paraffin oil (Protocol 1)
- 37°C, 5% CO_2 incubator

- Holding pipette (Section 2.4.2)
- Injection needle (see Part C, steps 2–4 below)
- Pipette puller
- Silicone rubber sheet
- Sharp scalpel blades
- Tygon tubing
- Glass syringe
- Gelatinized Petri dishes (see *Protocol 7*)
- Centrifuge
- Ethanol

A. Setting up matings

1   Day 0: Set up matings between C57BL/6 males and females. For better mating efficiencies females in oestrus may be selected by examining for the appearance of a pink and swollen vagina.

2   Day 1: Identify females that have mated by checking for copulation plugs and place them in a separate cage for later collection of 3.5-day blastocysts (day 4). Set up additional matings between F_1 (C57BL/6 × CBA) females and vasectomized males for the production of pseudopregnant females.

3   Day 2: Check the F_1 females for plugs and place those that have mated in a separate cage. They will be used on day 4 (at 2.5 days' pseudopregnancy) as recipients for the injected blastocysts.

B. Recovery of blastocysts

1   On day 4, kill the C57BL/6 females (collected on day 1) and dissect out both uterine horns using two incisions: one next to the oviduct and the second at the distal end junction of the two uterine horns.

2   Use a 1 ml syringe (25-G needle) filled with Hepes-buffered ESC medium to flush the blastocysts out of the uteri into a 6 cm tissue culture dish.

3   Collect the blastocysts under a stereo-microscope using a mouth-controlled, heat-drawn Pasteur pipette, wash in ESC medium and transfer into microdrop cultures overlaid with light paraffin oil. Store in a 5% CO_2, 37°C incubator. (At this stage, some blastocysts may not be fully expanded; they will do so later during the day.)

C. Preparation of the holding and injection pipettes

The holding pipette is essentially the same as the one used for pronuclear injections (see Section 2.4.2), except that a bend of approximately 30° is introduced 2–3 mm from its end, using a microforge.

1   The injection needle is used to collect individual ESC and introduce them into the blastocoel cavity. Use thin-walled borosillicate capillaries and a pipette puller to produce needles that have a relatively long section at the appropriate internal diameter, which should be slightly larger than the ESC (18–20 μm).

Protocol 13 continued

2 Place the pulled needle on a transparent silicone rubber sheet under a stereomicroscope and, using a sharp scalpel blade, snap it at the region of the appropriate diameter to create a sharp bevelled point.

3 Use the microforge to introduce a bend of approximately 30° close to the end of the injection needle, take care to keep the bevel facing to the side.

D. Setting up the injection chamber

1 The microscopes and micromanipulators described for pronuclear injections (see Section 2.1) are also suitable for blastocyst injections. Perform blastocyst injections in a lid of a 3.5 cm tissue culture dish semi-filled with Hepes-buffered ESC medium and overlaid with light paraffin oil. (Petri dish lids are conveniently shallow to allow free movement of the holding and injection pipettes.)

2 Set up the holding pipette as for pronuclear injections (see Sections 2.4.2 and 2.6). Fill the injection needle with light paraffin oil and connect through a length of Tygon tubing to a glass syringe controlled by a sensitive micrometer head.

3 Lower the holding and injection pipettes into the injection dish and make appropriate adjustments to position their ends parallel to the bottom of the dish.

E. Preparation of ESC for blastocyst injections

1 Between 3 and 4 h before harvesting, feed a subconfluent ESC-containing dish (3.5 or 6 cm) with fresh ESC medium.

2 Wash the plate twice with PBS, add trypsin–EDTA solution and place it back in the incubator for 3–5 min.

3 Add 3–4 ml of ESC medium and pipette vigorously to dissociate colonies into a single-cell suspension. Plate the suspension on a gelatinized dish (of the same size or larger).

4 Transfer the dish into the incubator for 35–40 min (no longer). (During this period, feeder cells attach strongly to the surface of the gelatinized dish, while most of the viable ESC only begin to adhere. Carefully, aspirate and discard the medium containing the non-adherent (non-viable) cells. Add 5 ml of ESC medium to the dish and suspend the loosely adhering ES cells by pipetting.

5 Centrifuge the suspension at 200 g for 5 min at RT and resuspend the cell pellet in 1–2 ml ESC medium which has been pre-cooled to 8 °C. Transfer a few μl into the injection dish and store the rest of the suspension at 8 °C.

F. Blastocyst injection

1 Transfer 10–20 blastocysts into the injection dish.

2 Collect 10–15 ESC in the injection needle using phase-contrast optics at 200 × magnification. Select the round cells that have a light yellow colour; cells that appear dark are dead or dying.

Protocol 13 continued

3   Use the holding pipette to pick up a blastocyst and focus on an appropriate point for injection. (These are usually points of intercellular junctions between adjacent trophectoderm cells.)

4   Bring the point of entry and the injection needle into the same focal plane. Push the needle into the blastocoel cavity with a steady and smooth movement. (Too slow a movement may result in blastocyst collapse making further penetration of the needle and injection impossible.

5   Release 10–15 ESC by carefully applying positive pressure. Note that after injection the blastocyst is seen to collapse.

6   After having injected all the blastocysts, place them back in the incubator. Check 1 h later that they have started to re-expand and that injected cells can be observed in the blastocoel, some of them being attached to the inner cell mass.

G. Transferring blastocysts into the uteri of pseudopregnant females

1   Anaesthetize the recipient female by an intraperitoneal injection of 0.3 ml of Hypnorm/Hypnovell/ddH$_2$O mix.

2   Swab the back of the mouse with ethanol and make a small midline incision in the skin at approximately the level of the last rib. Locate the position of the ovary, indicate by a pink structure seen through the body wall.

3   Make a small incision through the body wall and use blunt forceps to grasp and gently pull out the fat pad which is attached to the ovary, thereby exposing the uterus.

4   Use a 25-gauge needle to make a hole in the uterus close to the oviduct end.

5   Use a finely drawn Pasteur pipette to transfer six or seven blastocysts into the uterus through the hole created by the needle.

6   Gently push the uterus back into the peritoneal cavity and continue with the opposite uterine horn.

7   When the operation is completed, staple together the edges of the skin where the incision was made using one or two wound clips.

8   Keep the operated females warm until they have recovered from the anaesthetic. They should give birth 17–18 days later.

9   Identify chimeric newborns at around 7 days after birth by the presence of the agouti coat colour (derived from the ESC) on the black (C57BL/6) background. Since ESC lines with a male karyotype are more often used, expect a distortion of the sex ratio (towards males) in chimeric mice.

10   Cross male chimeras with C57BL/6 females to obtain ESC-derived progeny (agouti coat colour). Confirm germline transmission of the targeted allele, which is expected in 50% of the agouti progeny, by DNA analysis of mouse tail fragments.

11   Cross mice heterozygous for the targeted allele to obtain homozygous knockout mice.

(a)

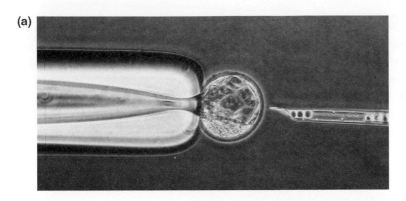

(b)

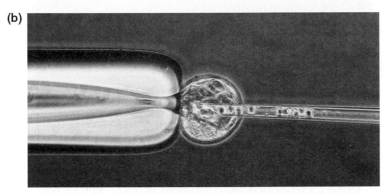

(c)

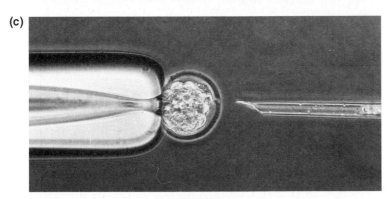

Figure 5 Injection of ESC into the blastocoel cavity of a mouse blastocyst. (a) The blastocyst is immobilized on the holding pipette and the needle containing the ESC is brought into focus. (b) ESC are expelled in the blastocoel cavity. (c) The blastocyst is starting to collapse after the ESC have been injected.

4 Conditional gene targeting

Conditional gene targeting denotes gene modifications that are restricted to a certain tissue or cell type, or to a specific developmental stage of the organism (see ref. 30). The site-specific recombinase Cre (*c*auses *re*combination) of the bacteriophage P1 has been widely used with high recombination efficiencies in

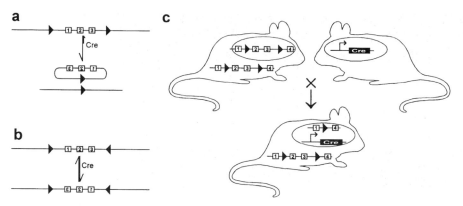

Figure 6 Cre-mediated recombination. Recombination between two loxP sites with the same orientation (a) causes excision of the intervening segment as a circular molecule, while recombination between two loxP sites with opposite orientation (b) causes inversion of the intervening segment. The Cre/loxP site-specific recombination system can be adapted for *in vivo* gene modifications in mice (c).

murine systems. Cre excises a DNA segment flanked by two tandemly repeated recognition sites, 34 bp in size, which are called loxP (*locus* of *x*-ing-over of *P*1) sites, as a circular molecule, whereas it inverts a DNA segment flanked by two loxP sites with opposite orientation (see *Figure 6*). Recombination between two loxP sites is carried out in a precise way that leaves behind one fully functional loxP site (*Figure 6*). The application of the Cre/loxP recombination system for conditional gene targeting requires the generation and crossing of two mouse lines: one carrying the gene of interest in a form that can be modified by Cre-mediated recombination and another expressing *cre* in certain tissues or cell types, either constitutively or following induction by exogenous stimuli (*Figure 6*).

4.1 Production of conditionally targeted alleles

Gene constructs that can be modified by Cre are usually introduced into the mouse genome by gene targeting (see Section 3). Therefore, the rules of conventional gene targeting also apply in this case. But, in general, the aim of gene targeting for the production of Cre-modifiable lines is to introduce loxP sites at specific and predefined places into the gene of interest, so that Cre-mediated recombination between these loxP sites will have the desired result. Alternatively, the Cre-modifiable construct can be introduced into mice by conventional transgenesis (see Section 2). However, this method often requires single-copy transgenes, while complications of position-effect variegation or ectopic and/or non-regulated expression of the transgene might turn out to be a serious impediment (see Section 2.7).

Conditional Cre-mediated gene modifications may include all kinds of gene modifications that are made possible with the use of the Cre/loxP system, but they usually aim at the reactivation, inactivation, or replacement of a predefined gene. All three types of conditional gene modifications have been successfully applied in mice (31–34). For conditional gene reactivation, the gene is initially inactivated by the introduction of loxP-flanked inhibitory sequences into an area important for gene expression. Such areas may be a specific intron or a site between the transcription initiation and the ATG codon of the gene of interest. Removal of the inhibitory sequences by Cre-mediated recombination should result in a fully functional gene (*Figure 7*). By crossing to a *cre*-expressing line, the inhibitory sequences are removed only in those tissues or only after those developmental stages in which *cre* is expressed, allowing the conditional expression (reactivation) of the gene. For conditional gene inactivation, initially, a functionally essential part of the gene or the whole gene is flanked by two loxP sites, in a way that does not influence its expression. This is carried out in two steps: in the first step, the selection marker flanked by two loxP sites with the same orientation is placed into the gene of interest. A third loxP is placed with the same orientation and at such a distance from the loxP-flanked selection marker so that an essential part of the gene is included between them. In the second step, the selection marker is removed by transient *cre* expression. The clones that will have undergone recombination only between the two loxP sites flanking the selection marker are selected (*Figure 7*). For conditional reversible gene replacement, a new gene (or a new part of the gene) is placed with the opposite orientation after the gene or the part of the gene that is going to be replaced.

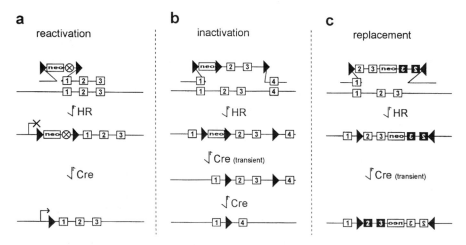

Figure 7 Strategies for *in vivo* gene modifications with the Cre/loxP site-specific recombination system successfully applied in mice. A predefined gene can be reactivated (a) after Cre-mediated excision of loxP-flanked inhibitory sequences that were previously introduced by homologous recombination (HR). Alternatively, a gene can be inactivated (b) after Cre-mediated excision of a loxP-flanked important part of the coding sequence. Flanking with loxP sites of such parts of a gene is carried out in two steps. Initially, a loxP-flanked selection marker and a third loxP site are introduced at certain points into the gene by HR. Subsequently the selection marker is selectively excised by transient Cre expression and selection for the desired product among all possible recombination products. Finally, a gene or a part of a gene can be replaced (c) with the use of two inverted loxP sites. After transient Cre expression half the loxP-flanked targets are expected to have the inverted configuration.

Two loxP sites with opposite orientation flank the two genes or the two parts of the gene. Transient expression of *cre* causes inversion of the loxP-flanked segment, thus replacing the gene (*Figure 7*).

4.2 Production of *cre-expressing transgenic lines*

4.2.1 Production of *cre*-expressing transgenic mice

Both conventional transgenesis (see Section 2) and targeted transgenesis (knockin approach) have been used for generating *cre*-expressing transgenic lines. Cell-specific expression of *cre* can be achieved by certain promoter/enhancer combinations. Targeted insertion of *cre* under the control of an endogenous promoter with a given expression pattern, ensures the correct and restricted expression of *cre*, but in some cases it has been associated with low *cre*-expression and hence recombination efficiency. The choice of promoter/enhancer combination should also be based on developmental timing of expression, since even transient *cre*-expression during development or in progenitor-cell types would result in generalized excision in the adult.

4.2.2 Production of transgenic mice expressing inducible cre

Inducible *cre* expression is extremely useful in the analysis of the role of certain genes in adult animals, because it offers the opportunity of gene modification after the completion of the normal development of the organism or of its adaptive cell systems, such as the neuronal and the immune system. While constitutive cell-specific *cre* expression can be achieved with the appropriate combination of natural or engineered promoters and enhancers, temporal control of Cre-mediated recombination can be exercised by regulating either the transcriptional or the post-translational level of *cre* expression. The simplest method for the inducible *cre* expression can be achieved with one of the systems that regulate the function of engineered promoters via specific transactivator proteins, which in turn are regulated by an inducer molecule. Expressing the transactivator in a cell-specific manner and *cre* under the control of such engineered promoters, the time of *cre* expression can be precisely controlled by the administration of the inducer. Both tetracycline and ecdysone have been successfully used in transgenic mice (36, 37). Post-translational control can be achieved by fusing Cre with a mutated ligand-binding domain (LBD) of steroid hormones that only bind synthetically produced steroid hormone analogues. The fusion protein can be expressed under the control of a constitutive cell-specific promoter, but it remains the cytoplasm. Administration of the synthetic analogue causes the translocation of the fusion protein to the nucleus where Cre gains access to loxP-flanked DNA targets (38–40).

4.2.3 Functional characterization of cre-*expressing transgenic lines*

Cre-mediated recombination can be assessed either at the DNA level (detection of recombined alleles by PCR or Southern analysis) or by the activation of a reporter gene. Detection of Cre protein (or mRNA) is not a reliable criterion because Cre-mediated recombination is a fairly permanent event compared to *cre* expression, and a given cell type that is not producing Cre mRNA or protein at the time of examination might have undergone Cre-mediated recombination earlier. Two key parameters of the functionality of any Cre-line are the efficiency and specificity of recombination. Analysis at the DNA level is very informative about the percentage of cells that have undergone recombination (efficiency), but for the identification of such cells (specificity), especially in solid tissues, direct visualization is preferred. This is feasible with the use of a reporter protein, such as β-galactosidase (β-gal), green fluorescent protein (GFP), or placental alkaline phosphatase (PLAP). A reporter line carries the gene encoding the reporter protein in an inactive form, which, however, can be reactivated by Cre-mediated recombination. This is usually achieved by the insertion of loxP-flanked inhibitory sequences between a ubiquitously expressed promoter and the reporter gene. By crossing the reporter line with a Cre-line the inhibitory sequences are removed only in *cre*-expressing cells, thus allowing their identification. We have routinely used an actin–*lacZ* transgenic line (41) and a knockin *lacZ* into the ROSA26 locus as reporter lines (42). The activity of *lacZ* can be assessed both in tissue sections (1) or by flow cytometry (43).

Protocol 14

Assessment of *lacZ* activity by cytochemistry

Equipment and reagents

- Fixative: 0.2% paraformaldehyde in 0.1 Pipes buffer (Sigma) pH 6.9, 2 mM $MgCl_2$, 5 mM EGTA
- Detergent rinse: 0.1 M phosphate buffer pH 7.3, 2 mM $MgCl_2$, 0.01% sodium deoxycholate, 0.02% Nonidet P-40 (NP-40, Sigma)
- X-gal (5-bromo-4-chloro-3-indolyl-β-D-galactopyranoside, Promega V3941)
- Staining solution: 0.1 M phosphate buffer pH 7.3, 2 mM $MgCl_2$, 0.1% sodium deoxycholate, 0.2% IGEPAL CA-630 (Sigma), 5 mM potassium ferricyanide, 5 mM potassium ferrocyanide, 1 mg/ml X-gal
- PBS plus 2 mM $MgCl_2$
- Cryotome
- OCT compound (BDH)
- Gelatin-coated slides
- Distilled water
- 1% (w/v) eosin in H_2O (Merck)
- Ethanol series: 50%, 70%, 100%
- Xylene
- DPX mountant (BDH)
- Bright-field optics
- Dry ice

Method

1 Embed freshly isolated material in OCT and freeze on dry ice.

2 Section the sample on a cryotome on to gelatin-coated slides (> 20 μm thick sections are preferred).

3 Fix sections in the 0.2% paraformaldehyde solution (fixative) for 10 min at 4 °C.

Protocol 14 continued

4 Rinse in PBS with 2 mM MgCl$_2$, followed by a 10 min wash in the same solution at 4°C.

5 Place in the detergent rinse for 10 min at 4°C.

6 Stain at 37°C in the dark. Note that the colour will become apparent in 2–24 h. For extended staining periods ($>$ 2 h), add Tris, pH 7.3, to the staining solution at a 20-mM final concentration for the best results. (This pH is optimum for the bacterial enzyme, while activity of the endogenous enzyme is kept to a minimum.)

7 Wash twice in PBS with 2 mM MgCl$_2$ at room temperature for 5 min each time.

8 Rinse in distilled water.

9 Counterstain with eosin.

10 Dehydrate through ethanol (5 min each in 50%, 70%, and 100% ethanol).

11 Clear in xylene and mount in DPX.

Protocol 15

Assessment of *lacZ* activity by flow cytometry

Equipment and reagents

- FACS buffer: PBS with 3% FCS, 0.1% sodium azide
- FDG (fluorescein digalactoside, Molecular Probes, F-1179): 20 mM stock solution in 8:1:1 (v/v) H$_2$O/DMSO/ethanol

- Distilled water
- Propidium iodide: 50 μM 10 $\times$ stock solution
- Flow cytometer

Method

1 Dilute the FDG stock solution 10-fold into sterile deionized water at 37°C (2 mM final concentration).

2 Mix the diluted FDG solution with an equal volume of the cell suspension in FACS buffer (approximately 10^6 cells per sample) and incubate for 1 min at 37°C. (The resulting hypotonic solution will permeabilize the cells, allowing the FDG to enter.)

3 After 1 min, dilute the suspension at least 10-fold with ice-colds FACS buffer.

4 Maintain the samples at 4°C on ice for 30–60 min until analysis. (At 4°C leakage of the fluorescent product from and into the cells is reduced.)

5 Analyse the samples by flow cytometry. Note that due to the severe nature of this loading method, it is recommended that propidium iodide (5 μM final concentration) is used for the exclusion of dead cells.

NB: The use of cell-specific markers before (preferably) or after the FDG staining allows the identification and classification of the Cre-mediated recombination-positive cells.

Although there is no direct evidence from our laboratory for intraline variability of *cre*-expression and recombination efficiency, this is a parameter that should always be examined in each Cre-line. Furthermore, the recombination efficiency of a certain Cre-line should not be considered invariable for different loxP-flanked target-genes and should be assessed for each target gene separately. Lists of currently available Cre-lines can be found at the following World Wide Web sites:

- **http://www.mshri.on.ca/nagy/cre.htm**—This web site is maintained by Nagy's lab. (Samuel Lunenfeld Research Institute, Mount Sinai Hospital, 600 University Avenue, Toronto, Ontario, Canada) and contains an excellent list of almost all Cre-lines that have been made or are currently being made.

- **http://www.pasteur.gr/molgen/animal_resources.htm**—This web site is maintained by our former lab. (former Lab. of Molecular Genetics, Hellenic Pasteur Institute, 127 Vas. Sofias Avenue, 115–21 Athens, Greece, now moved to Institute of Immunology, BSRC 'Al. Fleming', 14-16 Al. Fleming Str., Vari 166-72, Greece, http:/www.fleming.gr) and contains a detailed description of all the Cre-lines that have been generated by our lab.

- **gttp://jaxmice.jax.org/jaxmicedb/html/transgene.shtml**—This is the official web site of The Jackson Laboratory (600 Main Street, Bar Harbor, Maine 04609, USA) listing all the Cre-lines that can be purchased.

5 Transgenic and knockout mice in cytokine research

Transgenic and knockout systems offer clear advantages over cellular systems in the analysis of the functional potency of factors participating in complex multicellular processes. This is especially true when pleiotropic and redundant activities of factors are studied. For example, fine analysis of cytokine function-ing in the immune system necessitates the use of experimental settings where faithful measurement of *in vivo* reactivities, due in the presence or absence of a certain cytokine, may be easily performed.

Overexpressing or knocking-out cytokine and cytokine receptor genes in transgenic mice is currently providing much insight into the contribution of those factors to the maintenance of homeostasis or the triggering of disease in the course of immune responses. Further understanding of cytokine function-ing should come mainly through studies addressing the susceptibility or resist-ance of such 'mutant' mice to infectious or genetic disease. Current advances in the 'genetic engineering of the mouse', including the tissue-specific activation or inactivation of gene expression combined with developing technologies for switching gene expression on and off at will, provide experimental settings unprecedented in their potential to offer answers to long-standing questions or even to inspire currently unthought of questions.

Acknowledgements

Work in the author's laboratory is currently supported by the Hellenic Secretariat for Research and Technology and by European Commission grants BIO-CT96-0077 and BIO-CT96-0174.

References

1. Hogan, B., Beddington, R., Costantini, F., and Lacy, E. (1994). *Manipulating the mouse embryo: a laboratory manual (2nd edn).* Cold Spring Harbor Press, New York.
2. P. M. Wasserman and M. L. DePamphilis (ed.) (1993). *Methods in enzymology: guide to techniques in mouse development Vol. 225. Academic Press, London.*
3. Taketo, M., Schroeder, A. C., Mobraaten, L. E., Gunning, K. B., Hanten, G., Fox, R. R., Roderick, T. H., Stewart, C. L., Lilly, F., Hansen, C. T., and Overbeek, P. A. (1991). *Proc. Natl. Acad. Sci. USA*, **88**, 2065.
4. Brinster, R. L., Chen, H. Y., Trumbauer, M. E., Yagle, M. K., and Palmiter, R. D. (1985). *Proc. Natl. Acad. Sci. USA*, **82**, 4438.
5. Towens, T. M., Lingrel, J. B., Chen, H. Y., Brinster, R. L., and Palmiter, R. D. (1985). *EMBO J.*, **4**, 1715.
6. Brinster, R. L., Allen, J. M., Behringer, R. R., Gelinas, R. E., and Palmiter, R. D. (1988). *Proc. Natl. Acad. Sci. USA*, **85**, 836.
7. Sambrook, J., Fritch, E. F., and Maniatis, T. (ed.) (1989). *Molecular cloning, a laboratory manual* (2nd edn). Cold Spring Harbor Laboratory Press, NY.
8. Schedl, A., Larin, Z., Montoliu, L., Ties, E., Kelsey, G., Lehrach, H., and Schutz, G. (1993). *Nucleic Acids Res.*, **21**, 4783.
9. Strouboulis, J., Dillon, N., and Grosveld, F. (1992). *Genes Dev.*, **6**, 1857.
10. Mann, J. R. (1993). In *Methods in enzymology*, Vol. 225 (ed. P. M. Wasserman, and M. L. DePamphilis), pp. 782. Academic Press, London.
11. Probert, L., Keffer, J., Corbella, P., Cazlaris, H., Patsavoudi, E., Stephens, S., Kaslaris, E., Kioussis, D., and Kollias, G. (1993). *J. Immunol.*, **151**, 1894.
12. Grosveld, F., Blom van Assendelft, G. B., Greaves, D. R., and Kollias, G. (1987). *Cell*, **51**, 975.
13. Kollias, G. and Grosveld, F. (1992). In *Transgenic animals* (ed. F. Grosveld and G. Kollias), pp. 79–98. Academic Press, London.
14. Evans, M. J. and Kaufman, M. H. (1981). *Nature*, **292**, 154.
15. Bradley, A., Evans, M. J., Kaufman, M. H., Robertson, E. J. (1984). *Nature*, **309**, 255.
16. Robertson, E. J. (1987). In *Teratocarcinomas and embryonic stem cells: a practical approach* (ed. E. J. Robertson), pp. 71–112. IRL Press, Oxford.
17. Williams, R. L., Hilton, D. J., Pease, S., Wilson, T. A., Stewart, C. L., Gearing, D. P., Wagner, E. F., Metcalf, D., Nicola, N. A., and Gough, N. M. (1988). *Nature*, **336**, 684.
18. Smith, A. G., Heath, J. K., Donaldson, D. D., Wong, G. G., Moreau, J., Stahl, M., and Rogers, D. (1988). *Nature*, **336**, 688.
19. te-Riele, H., Maandag, E. R., and Berns, A. (1992). *proc. Natl. Acad. Sci. USA*, **89**, 5128.
20. Mansour, S. L., Thomas, K. R>, and Capecchi, M. R. (1988). *Nature*, **336**, 348.
21. Zimmer, A. and Gruss, P. (1989). *Nature*, **338**, 150.
22. Reid, L. H., Shesely, E. G., Kim, H.-S., and Smithies, O. (1991). *Mol. Cell. Biol.*, **11**, 2769.
23. Valancius, V. and Smithies, O. (1991). *Mol. Cell. Biol.*, **11**, 1402.
24. Hasty, P., Ramirez-Solis, R., Krumlauf, R., and Bradley, A. (1991). *Naturem* 350, 243.
25. Wu, H., Liu, X., and Jaenisch, R. (1994). *Proc. Natl. Acad. Sci. USA*, **91**, 2819.
26. Ramirez-Solis, R., Rivera-Perez, J., Wallace, J. D., Wims, M., Zheng, H., and Bradley, A. (1992). *Anal. Biochem.*, **201**, 331.

27. Stewart, C. L. (1993). In *Methods in enzymology*, Vol. 225 (ed. P. M. Wasserman, and M. L. DePamphilis), pp. 843–8. Academic Press, London.

28. Wood, S. A., Pascoe, W. S., Schmidt, C., Kemler, R., Evans, M., and Allen, N. (1993). *Proc. Natl. Acad. Sci. USA*, **90**, 4582.

29. Schwarzberg, P. L., Goff, S. P., and Robertson, E. J. (1989). *Science*, **246**, 799.

30. Torres, R. M. and Kuhn, R. (1997). *Laboratory protocols for conditional gene targeting*, pp. 1–126, Oxford University Press, Oxford.

31. Gu, H., Marth, J. D., Orban, P. C., Mossmann, H., and Rajewsky, K. (1994). *Science*, **265**, 103.

32. Tsien, J. Z., Chen, D. F., Gerber, D., Tom, C., Mercer, E. H., Anderson, D. J., Mayford, M., Kandel, E. R., and Tonegawa, S. (1996). *Cell*, **87**, 1317.

33. Akagi, K., Sandig, V., Vooijs, M., Van der Valk, M., Giovannini, M., Strauss, M., and Berns, A. (1997). *Nucleic Acids Res.*, **25**, 1766.

34. Lam, K. P. and Rajewsky, K. (1998). *Proc. Natl. Acad. Sci. USA*, **95**, 13171.

35. Kuhn, R., Schwenk, F., Aguet, M., and Rajewsky, K. (1995). *Science*, **269**, 1427.

36. Furth, P. A., St Onge, L., Boger, H., Gruss, P., Gossen, M., Kistner, A., Bujard, H., and Hennighausen, L. (1994). *Proc. Natl. Acad. Sci. USA*, **91**, 9302.

37. No, D., Yao, T. P., and Evans, R. M. (1996). *Proc. Natl. Acad. Sci. USA*, **93**, 3346.

38. Zhang, Y., Riesterer, C., Ayrall, A. M., Sablitzky, F., Littlewood, T. D., and Reth, M. (1996). *Nucleic Acids Res.*, **24**, 543.

39. Feil, R., Brocard, J., Mascrez, B., LeMeur, M., Metzger, D., and Chambon, P. (1996). *Proc. Natl. Acad. Sci. USA*, **93**, 10998.

40. Kellendonk, C., Tronche, F., Monaghan, A. P., Angrand, P. O., Stewart, F., and Schutz, G. (1996). *Nucleic Acids Res*, **24**, 1404.

41. Akagi, K., Sandig, V., Vooijs, M., Van der Valk, M., Giovannini, M., Strauss, M., and Berns, A. (1997). *Nucleic Acids Res.*, **25**, 1766.

42. Soriano, P. (1999). *Nature Genet.*, **21**, 70.

43. Nolan, G. P., Fiering, S., Nicolas, J. F., and Herzenberg, L. A. (1988). *Proc. Natl. Acad. Sci. USA*, **85**, 2603.

List of suppliers

Accurate Chemical and Scientific Corp., 300 Shames Drive, Westbury, NY 11590, USA

Advanced Biotechnologies Ltd, Units B1-B2, Longmead Business Centre, Blenheim Road, Epsom, Surrey KT19 9QQ
Tel: 01372 723456
Fax: 01372 741414
URL: http//www.abgene.com

Alexis, 3 Moorbridge Court, Moorbridge Road East, Bingham, Nottingham NG13 8QG
Tel: 01949 836111
Fax: 01949 936222
URL: http//www.alexis-corp.com

Amersham Pharmacia Biotech UK Ltd, Amersham Place, Little Chalfont, Buckinghamshire HP7 9NA, UK (see also Nycomed Amersham Imaging UK; Pharmacia)
Tel: 0800 515313   Fax: 0800 616927
URL: http//www.apbiotech.com/
Amersham Pharmacia Biotech, Björkgatan 30, 75184 Uppsala, Sweden

Amersham Corp., 2636 South Clearbrook Drive, Arlington Heights, IL 60005, USA

Anderman and Co. Ltd, 145 London Road, Kingston-upon-Thames, Surrey KT2 6NH, UK
Tel: 0181 5410035
Fax: 0181 5410623

Applied Immune Sciences Inc., 5301 Patrick Henry Drive, Santa Clara, CA 95054-1114, USA

Applied Scientific, San Francisco, CA 944080, USA

Arnold R. Horwell, 73 Maygrove Rd, West Hampstead, London NW6 2BP, UK

Associates of Cape Cod, 704 Main Street, Falmouth, MA 02540, USA

ATCC (American Type Culture Collection), 10801 University Boulevard, Manassas, VA 20110–2209, USA

Autogen Bioclear, Holditch Farm, Mile Elm, Calne, Wiltshire SN11 0PY
Tel: 01249 819008
Fax: 01249 817266
URL: http://www.autogen-bioclear.co.uk

Baxter Healthcare Ltd., Wallingford Road, Compton, Nr Newbury, Berkshire RG20 7QW
Tel: 01635 206 000   Fax: 01635 206 115
URL: http://www.baxter.com

Baxter Healthcare Corporation, Deerfield, IL 60015, USA

BDH Laboratory Supplies, Poole, Dorset BH15 1TD
URL: http://www.bdh.com

Beckman Coulter (UK) Ltd, Oakley Court, Kingsmead Business Park, London Road, High Wycombe, Buckinghamshire HP11 1JU, UK
Tel: 01494 441181   Fax: 01494 447558
URL: http://www.beckman.com/
Beckman Coulter Inc., 4300 N. Harbor Boulevard, PO Box 3100, Fullerton, CA 92834–3100, USA
Tel: 001 714 8714848
Fax: 001 714 7738283
URL: http://www.beckman.com/

Beckman Instruments, Progress Rd, Sands Industrial Estate, High Wycombe, Buckinghamshire HP12 4JL, UK; PO Box 3100, 2500 Harbor Boulevard, Fullerton, CA 92634, USA

Becton Dickinson and Co., 21 Between Towns Road, Cowley, Oxford OX4 3LY, UK
Tel: 01865 748844   Fax: 01865 781627
URL: http://www.bd.com/
Becton Dickinson and Co., 1 Becton Drive, Franklin Lakes, NJ 07417–1883, USA
Tel: 001 201 8476800
URL: http://www.bd.com/

Bellco Glassware (distributed in the UK by Philip Harris Scientific); PO Box 340, Edrudo Rd, Vineland, NJ 08360, USA

Bio 101 Inc., c/o Anachem Ltd, Anachem House, 20 Charles Street, Luton, Bedfordshire LU2 0EB, UK
Tel: 01582 456666
Fax: 01582 391768
URL: http://www.anachem.co.uk/
Bio 101 Inc., PO Box 2284, La Jolla, CA 92038–2284, USA
Tel: 001 760 5987299
Fax: 001 760 5980116
URL: http://www.bio101.com/

Biochrom KG, Leonorenstr. 2-6, D-12247, Berlin, Germany

Bio-Rad Laboratories Ltd, Bio-Rad House, Maylands Avenue, Hemel Hempstead, Hertfordshire HP2 7TD, UK
Tel: 0181 3282000
Fax: 0181 3282550
URL: http://www.bio-rad.com/
Bio-Rad Laboratories Ltd, Division Headquarters, 1000 Alfred Noble Drive, Hercules, CA 94547, USA
Tel: 001 510 7247000
Fax: 001 510 7415817
URL: http://www.bio-rad.com/

Biognostik, c/o Chemicon International Ltd, 2 Admiral House, Cardinal Way, Harrow NA3 5UT, UK; 28835 Single Oak Drive, Temecula, CA 92590, USA

BioWhittaker, BioWhittaker House, 1 Ashville Way, Wokingham RG41 2PL
Tel: 0118 979 5234
Fax: 0800 731 3498
URL: http://www.biowhittaker.com

Boehringer-Mannheim (see Roche Diagnostics)

Brickmann Instruments Inc., 1 Cantiague Rd, Westbury, NY 11590, USA

British Drug Houses (BDH) Ltd, Poole, Dorset, UK

Buck&Holm A/S, Marielundvej 36, DK-2730 Herlev, Denmark

Calbiochem, Boulevard Industrial Park, Padge Road, Beeston, Nottingham NG9 2JR

Campden Instruments, 185 Campden Hill Rd, London W8 1TH, UK

Cappelen Laboratory Technics (CAPP), Kallerupvej 26, PO Box 824, DK-5230 Odense M, Denmark

Carl Zeiss, D-7082, Oberkochen, Germany

CellPro, Inc., Suite 100, 22322-20th Avenue Southeast, Bothell, Washington 982021, USA
Tel: 001 206 485 7644
Fax: 001 206 485 4787

Clontech Laboratories, Unit 2, Intec 2, Wade Rd, Basingstoke, Hampshire RG24 8NE,UK; 1020 East Meadow Circle, Palo Alto, CA 94303, USA

Collaborative Research Inc., Two Oak Park, Bedford, MA 01730, USA

Corning Inc., Corning, NY 14831, USA

Costar/Nuclepore, 1 Alewife Center, Cambridge, MA 02140, USA

Coulter (see Beckman Coulter)

CP Instrument Co. Ltd, PO Box 22, Bishops Stortford, Hertfordshire CM23 3DX, UK
Tel: 01279 757711   Fax: 01279 755785
URL: http//:www.cpinstrument.co.uk/

Dako, Denmark House, Angel Drive, Ely, Cambridgeshire CB7 4ET, UK; 6392 Via Real, Carpinteria, CA 93013, USA

David Kopf, Tujunga, CA, USA

Difco Laboratories Ltd, PO Box 14B, Central Ave, West Moseley, Surrey, KT8 2SE, UK; PO Box 331058, Detroit, MI 48232–7058, USA

Dow Chemicals Company Ltd., Hydrocarbon, 1 Mount Street, London W1

Drummond Scientific Co., 500 Parkway, Broomal, PA 19008, USA

Dupont (UK) Ltd, Industrial Products Division, Wedgwood Way, Stevenage, Hertfordshire SG1 4QN, UK
Tel: 01438 734000
Fax: 01438 734382
URL: http://www.dupont.com/

Du Pont Co. (Biotechnology Systems Division), PO Box 80024, Wilmington, DE 19880–002, USA
Tel: 001 302 7741000
Fax: 001 302 7747321
URL: http://www.dupont.com/

Dynal (UK) Ltd., 10 Thursby Road, Croft Business Park, Bromborough, Wirral, Merseyside L62 3PW
Tel: 0151 346 1234
Fax: 0151 346 1223
techserve@dynal.u-net.com

Dynal (UK) Ltd., 11 Bassendale Road, Croft Business Park, Bromborough, Wirral CH62 3QL
Tel: 0800 731 9037
Fax: 0151 346 1223
URL: http://www.dynal.net

Dynatech, Daux Road, Billingshurst, West Sussex JRH14 9SJ

Eastman Chemical Co., 100 North Eastman Road, PO Box 511, Kingsport, TN 37662–5075, USA
Tel: 001 423 2292000
URL: http//:www.eastman.com/

Eppendorf, 2000 Hamburg 65–Postfach 650670, Germany

European Collection of Animal Cell Culture, Division of Biologics, PHLS Centre for Applied Microbiology and Research, Porton Down, Salisbury, Wiltshire SP4 0JG, UK

EY Labs, 107-127 N. Amphlett Blvd., San Mateo, California 94401, USA
Tel: 650 342 3296
Fax: 650 342 2648
URL: http//:www.eylabs.com

Falcon (Falcon is a registered trademark of Becton Dickinson and Co.)

Fisher Scientific UK Ltd, Bishop Meadow Road, Loughborough, Leicestershire LE11 5RG, UK
Tel: 01509 231166
Fax: 01509 231893
URL: http://www.fisher.co.uk/
Fisher Scientific, Fisher Research, 2761 Walnut Avenue, Tustin, CA 92780, USA
Tel: 001 714 6694600
Fax: 001 714 6691613
URL: http://www.fishersci.com/

Flow Laboratories, Woodcock Hill, Harefield Rd, Rickmansworth, Hertfordshire WD3 1PQ, UK

Fluka Chemicals Ltd (see Sigma–Aldrich)
Fluka, PO Box 2060, Milwaukee, WI 53201, USA
Tel: 001 414 2735013
Fax: 001 414 2734979
URL: http://www.sigma-aldrich.com/
Fluka Chemical Co. Ltd, PO Box 260, CH-9471, Buchs, Switzerland
Tel: 0041 81 7452828
Fax: 0041 81 7565449
URL: http://www.sigma-aldrich.com/

Fresenius Hemotechnology Inc., 110 Mason Circle, Suite A, Concord, CA 94520, USA; Bad Homburg, Germany

Gibco BRL (Life Technologies Ltd), Trident House, Renfrew Rd, Paisley PA3 4EF, UK; 3175 Staler Rd, Grand Island, NY 14072–0068, USA

Harleco, 480 Democrat Road, Gibbstown, New Jersey 08027, USA

Heto-Holten A/S, Gydevang 17–19, DK-3450 Allerød, Denmark

F. Hoffmann-La Roche Ltd., CH-4070 Basel, Switzerland
Tel: +41 61688 1111   Fax: +41 61691 9391
URL: http//:www.roche.com

Hybaid Ltd, Action Court, Ashford Road, Ashford, Middlesex TW15 1XB, UK
Tel: 01784 425000   Fax: 01784 248085
URL: http://www.hybaid.com/
Hybaid US, 8 East Forge Parkway, Franklin, MA 02038, USA
Tel: 001 508 5416918
Fax: 001 508 5413041
URL: http://www.hybaid.com/

HyClone Laboratories, 1725 South HyClone Road, Logan, UT 84321, USA
Tel: 001 435 7534584
Fax: 001 435 7534589
URL: http//:www.hyclone.com/

ICN Pharmaceuticals Ltd, 1 Elmwood, Chineham Business Park, Basingstoke, Hampshire RG 24 8WG, UK

IKA-Werke GmbH & Co., KG, Postfach 1263, D-79217 Staufen, Germany

Immunoprecipitin, Bethesda Research Laboratories, MD, USA

International Biotechnologies Inc., 25 Science Park, New Haven, CT 06535, USA

Intracel Ltd, Unit 4, Station Rd, Shepreth, Hertfordshire SG8 6PZ, UK

Invitrogen Corp., 1600 Faraday Avenue, Carlsbad, CA 92008, USA
Tel: 001 760 6037200
Fax: 001 760 6037201
URL: http://www.invitrogen.com/
Invitrogen BV, PO Box 2312, 9704 CH Groningen, The Netherlands
Tel: 00800 53455345
Fax: 00800 78907890
URL: http://www.invitrogen.com/

Jackson Laboratories, 872 West Baltimore Pike, PO Box 9, West Grove, Pennsylvania 19390, USA

Janke and Kunkel IKA-Labortechnik, Neumagenstrasse 27, 79219 Staufen, Germany

Kodak: Eastman Fine Chemicals, 343 State St, Rochester, NY, USA

LabTech International, 1 Acorn House, The Broyle, Ringmer, E. Sussex BN8 5NW
Tel: 01273 814 888
URL: http//:www.labtech.co.uk

Leica Mikrosysteme Vertrieb GmbH, Lilienthalstrasse 39-45, D-64625 Bensheim, Germany
Tel: +49 6251136 0
Fax: +49 6251136 155
URL: http//:www.leica.com

Life Technologies Ltd, PO Box 35, 3 Free Fountain Drive, Inchinnan Business Park, Paisley PA4 9RF, UK
Tel: 0800 269210   Fax: 0800 243485
URL: http://www.lifetech.com/
Life Technologies Inc., 9800 Medical Center Drive, Rockville, MD 20850, USA
Tel: 001 301 6108000
URL: http://www.lifetech.com/

MACS, Miltenyi Biotec Inc., Almac House, Church Lane, Bisley, Surrey GU24 9DR, UK; 251 Auburn Ravine Rd, Suite 208, Auburn. CA 95603, USA

Medicell International Ltd., 239 Liverpool Road, London N1 1LX
Tel: 020 7607 2295
Fax: 020 7700 4156

Merck, Sharp, & Dohme Research Laboratories, Neuroscience Research Centre, Terlings Park, Harlow, Essex CM20 2QR, UK
URL: http://www.msd-nrc.co.uk/
Merck Industries Inc., 5 Skyline Drive, Nawthorne, NY 10532, USA
Merck, Sharp, and Dohme GmbH, Lindenplatz 1, D-85540, Haar, Germany
URL: http://www.msd-deutschland.com/

MIAB, Box 97, S-74100 Knivsta, Uppsala, Sweden

Micro Instruments Ltd, 18 Hanborough Park, Long Hamborough, Witney, Oxon OX8 8LH, UK

Millipore (UK) Ltd, The Boulevard, Blackmoor Lane, Watford, Hertfordshire WD1 8YW, UK
Tel: 01923 816375
Fax: 01923 818297
URL: http://www.millipore.com/local/UKhtm/
Millipore Corp., PO Box 255, 80 Ashby Road, Bedford, MA 01730, USA
Tel: 001 800 6455476
Fax: 001 800 6455439
URL: http://www.millipore.com/

Miltenyi Biotec Ltd., Almac House, Church Lane, Bisley, Surrey GU24 9DR
Tel: 01483 799800   Fax: 01483 799811
macs@miltenyibiotec.co.uk

MJ Research, Inc., Waltham, MA 02451, U.S.A.
URL: http://www.mjresearch.com

Molecular Probes Europe BV, Poortgebouw, Rijnsburgerweg 10, 2333 AA Leiden, The Netherlands

Nalge Nunc International (distributed in the UK by Fisher Scientific UK)

Nalge Nunc, 75 Panorama Creek Drive, PO Box 20365, Rochester, NY 14602-0365, USA
Tel: 800 625 4327
Fax: 716 586 8987
URL: http://www.nalgecnunc.com

Nalgene (distributed in the UK by Techmate Ltd), 10 Bridgeturn Ave, Old Wolverton, Milton Keynes MK12 5QL, UK; 75 Panorama Creek Dr., PO Box 20365, Rochester, NY 14602–0365, USA

Narashige Scientific Instrument Laboratory, 9.28 Kasuya, 4 Chome Setagayaku, Tokyo, Japan

Nasco Co., 901 Janesville Ave, Ft Atkinson, WI 53538, USA

National Diagnostics, 305 Patton Drive, SW Atlanta, GA 30336, USA

NEN Life Sciences, 549 Albany St, Boston, MA 02118, USA

Neuro Probe Inc., 16008 Industrial Drive, Gaithersburg, MD 20877, USA

Neuroprobe, Cabin John, Madison, USA; Neuroprobe, 7621 Cabin Road, Bethesda, Maryland 20034, USA; Neuroprobe, Porvair Filtronics Ltd., Shepperton, Middx.

New England Biolabs (NBL) (distributed by CP Instruments UK Ltd); 32 Tozer Road, Beverley, MA 01915–5510, USA
Tel: 001 978 9275054

Nikon (UK) Ltd, Instrument Division, Haybrook, Halesfield 9, Telford, Shropshire T7 4EW, UK
Nikon Inc., 1300 Walt Whitman Road, Melville, NY 11747–3064, USA
Tel: 001 516 5474200
Fax: 001 516 5470299
URL: http://www.nikonusa.com/
Nikon Corp., Fuji Building, 2–3 Marunouchi 3-chome, Chiyoda-ku, Tokyo 100, Japan
Tel: 00813 32145311
Fax: 00813 32015856
URL: http://www.nikon.co.jp/main/index_e.htm/

Northumbria Biologicals Ltd., Nelson Industrial Estate, Cramlington, Northumberland NE23 9BL
Tel: 01670 732537
Fax: 01670 732992

Novagen Inc., 601 Science Drive, Madison, WI 53711, USA
URL: http://www.novagen.com

Nuclepore, Corning Costar, 1 The Valley Centre, Gordon Rd, High Wycombe, Buckinghamshire HP13 6EQ, UK; 45 Nagog Park, Acton, MA 01720, USA

Nycomed Amersham Imaging, Amersham Labs, White Lion Rd, Amersham, Buckinghamshire HP7 9LL, UK
Tel: 0800 558822 (or 01494 544000)
Fax: 0800 669933 (or 01494 542266)
URL: http//:www.amersham.co.uk/
Nycomed Amersham, 101 Carnegie Center, Princeton, NJ 08540, USA
Tel: 001 609 5146000
URL: http://www.amersham.co.uk/

Nycomed Pharma (distributed by Life Technologies (UK) Ltd; 3175 Staler Rd, Grand Island, NY 14072–0068, USA)
Nycomed Pharma, PO Box 4220 Torshov, N-0401, Oslo, Norway

Nyegaard Ltd, Mylen House, 11 Wagon Lane, Sheldon, Birmingham B26 3DU, UK (distributed in the USA by Accurate Chemical and Scientific Corporation)

Olympus Optical Co. (Europa) GmbH, Postfach 104908, Wendenstrasse 14–16, 2 Hamburg 1, Germany

PEBiosystems, Kelvin Close, Birchwood Science Park North, Warrington, Cheshire WA3 7PB
URL: http://www.appliedbiosystems.com

PE Biosystems, Division Headquarters, 850 Lincoln Centre Drive, Foster City, CA 94404, USA
URL: http//:www.appliedbiosystems.com

PBL Biomedical Laboratories, 100 Jersey Ave, Building D, New Brunswick, NJ 08901, USA

Perkin Elmer Ltd, Post Office Lane, Beaconsfield, Buckinghamshire HP9 1QA, UK
Tel: 01494 676161
URL: http//:www.perkin-elmer.com/
Perkin Elmer–Cetus (The Perkin–Elmer Corp.), 761 Main Ave, Norwalk, CT 0689, USA

Pharmacia Biosystems Ltd (Biotechnology Division), Davy Avenue, Knowlhill, Milton Keynes, Buckinghamshire MK5 8PH, UK (see also Amersham Pharmacia Biotech)
Tel: 01908 661101
Fax: 01908 690091
URL: http//www.eu.pnu.com/
Pharmacia Biotech Europe, Procordia EuroCentre, Rue de la Fuse-e 62, B-1130 Brussels, Belgium
Pharmacia LKB Biotechnology AB, Bjorngatan 30, S-75182 Uppsala, Sweden

PharMingen, 10975 Torreyana Road, San Diego, CA 92121, USA
Tel: 800 848 6227
Fax: 858 812 8888
URL: http//:www.pharmingen.com

Philip Harris Scientific, Lichfield, Staffordshire WS14 0EE, UK.

Pierce, 3747 N. Meridian Rd, PO Box 117, Rockfort, IL 61105, USA

Pierce & Warriner Ltd, 44, Upper Northgate St, Chester, Cheshire CH1 4EF, UK

Pillar Engineering Supplies Ltd, 103/109 Waldegrave Rd, Teddington, Middlesex TW11 8LL, UK

Polysciences Inc., 400 Valley Road, Warrington, PA 18976-2522, USA
Tel: 215 343 6485
URL: http//:www.polysciences.com

Promega UK Ltd, Delta House, Enterprise Rd, Chilworth Research Centre, Southampton SO16 7NS, UK
Tel: 0800 378994   Fax: 0800 181037
URL: http://www.promega.com/
Promega Corp., 2800 Woods Hollow Road, Madison, WI 53711–5399, USA
Tel: 001 608 2744330
Fax: 001 608 2772516
URL: http://www.promega.com/

Qiagen UK Ltd, Boundary Court, Gatwick Road, Crawley, West Sussex RH10 2AX, UK
Tel: 01293 422911   Fax: 01293 422922
URL: http://www.qiagen.com/
Qiagen Inc., 28159 Avenue Stanford, Valencia, CA 91355, USA
Tel: 001 800 4268157
Fax: 001 800 7182056
URL: http://www.qiagen.com/

R&D Systems, 4-10 The Quadrant, Barton Lane, Abingdon, Oxon OX14 3YS
Tel: 01235 529449   Fax: 01235 533420
URL: http://www.rndsystems.com

Roche Diagnostics Ltd, Bell Lane, Lewes, East Sussex BN7 1LG, UK
Tel: 0808 1009998 (or 01273 480044)
Fax: 0808 1001920 (or 01273 480266)
URL: http://www.roche.com/
Roche Diagnostics Corp., 9115 Hague Road, PO Box 50457, Indianapolis, IN 46256, USA
Tel: 001 317 8452358
Fax: 001 317 5762126
URL: http://www.roche.com/
Roche Diagnostics GmbH, Sandhoferstrasse 116, 68305 Mannheim, Germany
Tel: 0049 621 7594747
Fax: 0049 621 7594002
URL: http://www.roche.com/

Roussel Labs, Uxbridge, UB9 5HP
Tel: 01895 837597
Fax: 01732 584080
URL: http://www.aventis.com

Santa Cruz (distributed in UK by Autogen Bioclear)

Savant Instruments, Inc., 100 Colin Drive, Holbrook, NY 11741-4306, USA
Tel: 631 244 2929
Fax: 631 244 0606
URL: http://www.savec.com

Schleicher and Schuell Inc., Keene, NH 03431A, USA
Tel: 001 603 3572398
Schleicher & Schuell GmbH, Hahnestrasse 3, D-37586 Dassel, Germany

Schott Corp., 3 Odell Plaza,Yonkers, NY 10701, USA
Schott Glass, Mainz, Germany

Serva, Novex Experimental Technologies, 11040 Roselle Street, San Diego, CA 92121, USA.
Serva, Novex Electrophoresis GmbH, Brüningstrasse 50, Gebäude C 584, D-65929 Frankfurt am Main, Germany

Shandon Scientific Ltd, 93–96 Chadwick Road, Astmoor, Runcorn, Cheshire WA7 1PR, UK
Tel: 01928 566611
URL: http//www.shandon.com/

Sigma–Aldrich Co. Ltd, The Old Brickyard, New Road, Gillingham, Dorset SP8 4XT, UK
Tel: 0800 717181 (or 01747 822211)
Fax: 0800 378538 (or 01747 823779)
URL: http://www.sigma-aldrich.com/
Sigma Chemical Co., PO Box 14508, St Louis, MO 63178, USA
Tel: 001 314 7715765
Fax: 001 314 7715757
URL: http://www.sigma-aldrich.com/

Sorvall, 31 Pecks Lane, Newtown, CT 06470-2337, USA
URL: http://www.sorvall.com

Sorvall Du Pont Co., Biotechnology Division, PO Box 80022, Wilmington, DE 19880–0022, USA

Spectrum Laboratories (distributed in the UK by Pierce & Warriner Ltd); 23022 La Cadena Drive, Laguna Hills, CA 92653, USA
Spectrum Europe BV, PO Box 3262, 4800 DG Breda, The Netherlands

StemCell Technologies, Metachem Diagnostics Ltd., 29 Forest Road, Piddington, Northampton NN7 2DA
Tel: 01604 870370
Fax: 01604 870194
metachem@skynet.co.uk

Stratagene Ltd., Unit 140, Cambridge Innovation Centre, Milton Rd, Cambridge CB4 44FG, UK
Stratagene Inc., 11011 North Torrey Pines Road, La Jolla, CA 92037, USA
Tel: 001 858 5355400
URL: http://www.stratagene.com/
Stratagene Europe, Gebouw California, Hogehilweg 15, 1101 CB Amsterdam Zuidoost, The Netherlands
Tel: 00800 91009100
URL: http://www.stratagene.com/

Techmate Ltd, 10 Bridgeturn Ave., Old Wolverton, Milton Keynes MK12 5QL, UK.

3M, 3M House, PO Box 1, Bracknell, Berks. RG12 1JU
URL: http://www.3m.com

ThermoQuest Scientific Equipment Group Ltd., Unit 5, The Ringway Centre, Edison Road, Basingstoke, Hampshire. RG21 6YH
Tel: 01256 817282
Fax: 01256 817292

Tissue Culture Services, Boltolph Claydon, Bucks. MK18 2LR

Tommy Nielsen Handels, og Ingeniørfirma, Malervej 6, DK-6710 Esbjerg V, Denmark

UBI (distributed by TCS biologicals)

United States Biochemical (USB), PO Box 22400, Cleveland, OH 44122, USA
Tel: 001 216 4649277

Wallac Finland Oy, Mustionkatu 6, PL 10, 20101 Turku, Finnland

Waters Ltd, The Boulevard, Blackmoor Lane, Watford Hertfordshire, WD1 8YW, UK
Waters Corp., 34 Maple Street, Milford, MA 01757, USA

Wellcome Reagents, Langley Court, Beckenham, Kent BR3 3BS, UK

Wild Leitz UK Ltd, Davy Ave, Knowlhill, Milton Keynes, MK5 8LB, UK

Worthington Biochemical Corporation, Halls Road, Freehold, NJ 07728, USA

Zinsser Analytic, Eschborner Landstr. 135, D-60489 Frankfurt, Germany

Zymed, 458 Carlton Court, S. San Francisco, CA 94080, USA

Index